The Nature Center at W.W. Knight Preserve
29530 White Road
Perrysburg, OH 43551

The
Urban Cliff Revolution

The Urban Cliff Revolution

New Findings
on the Origins and Evolution
of Human Habitats

Douglas Larson, Uta Matthes, Peter E. Kelly,
Jeremy Lundholm, John Gerrath

Fitzhenry & Whiteside

Published in Canada by
   Fitzhenry & Whiteside, 195 Allstate Parkway, Markham, Ontario L3R 4T8

Published in the United States by
   Fitzhenry & Whiteside, 121 Harvard Avenue, Suite 2, Allston, Massachusetts 02134

www.fitzhenry.ca   godwit@fitzhenry.ca

10 9 8 7 6 5 4 3 2 1

National Library of Canada Cataloguing in Publication

The urban cliff revolution : new findings on the origins and evolution of human habitats /
Douglas Larson ... [et al.].

Includes bibliographical references and index.
ISBN 1-55041-848-3
1. Human settlements —History. I. Larson, Douglas W. (Douglas William), 1949-

GF101.U73 2004   307   C2004-902429-9

U.S. Publisher Cataloging-in-Publication Data
(Library of Congress Standards)

Larson, Douglas W. (Douglas William), 1949-
   The urban cliff revolution : new findings on the origins and evolution of human
habitats / Douglas Larson ; Uta Matthes ; Peter E. Kelly ; Jeremy Lundholm ; John
Gerrath. —1st ed.
[230] p. : ill., photos. ; cm.
Includes bibliographical references and index.
ISBN 1-55041-848-3 (pbk.)
1. Cliff ecology. I. Kelly, Peter E., 1963- II. Matthes, Uta, 1955- III. Lundholm, Jeremy. IV.
Gerrath, John. V. Title.
577 dc22   QH 541.5 .C62L27 2004

Fitzhenry & Whiteside acknowledges with thanks the Canada Council for the Arts, the Government of Canada through the Book Publishing Industry Development Program (BPIDP), and the Ontario Arts Council for their support of our publishing program.

Design by Karen Petherick
Cover design by Kerry Plumley
Top cover photo by Dr. Galen R. Frysinger
Printed in Canada

— ■ —

To the spirit of
Nikolay Ivanovich Vavilov
(1887–1943)
who sought large truths
and died for them

—■—

C O N T E N T S

CONTENTS

CONTENTS

THE FAR SIDE® By GARY LARSON

Stand-semi-erect comedy

Figure 1.
Cartoon image from the collection of Gary Larson. Used with permission.

— ■ —

We sometimes laugh and sometimes cringe at the familiar caricatures of ourselves as cave dwellers. We laugh because we *know* that they are not *us* (Figure 1). We cringe because we suspect they are. The nervousness is well founded. We may not be as different as we think. And, as this book argues, our sophisticated habitats may not be all that different from theirs.

This book puts forth a revolutionary new thesis: that rock outcrops and cliffs have played a central role in the origin, evolution, and development of the entire human habitat. It covers a time period of at least a million years and applies to all parts of the world. In the process, the book contrasts the way our world has been viewed by conventional thinking regarding evolution, history, and ecology with our new view, which we call the Urban Cliff Hypothesis. We will point out the ecological similarities between ancestral human habitats and modern ones and in the process provide new perspectives on what it means to be human.

We believe that the story we tell will change our understanding of the environment we have created for ourselves. Our hypothesis has implications in fields as diverse as architecture, landscape architecture, nature interpretation, and ecological restoration.

As for the way we have written the book, we have sought to stand in the new tradition of books by Stephen Jay Gould, E.O. Wilson, David Quammen, Ian Tattersall, and Wade Davis, writers who have improved the way scientific ideas are presented to the public. This type of book shows scientists to be curious people (sometimes in both senses of the description) who fumble around with vague ideas and apparently contradictory evidence well before they know how to interpret their findings. This approach allows us to reveal the way the scientific method has worked for us. Most peer-reviewed scientific literature makes the authors sound as if they knew the results in advance. But in fact most scientific research is not a linear journey from hypothesis to conclusions, and in this book we will not misrepresent what for us has been a sometimes circuitous voyage of discovery.

Chapter 1 presents the Urban Cliff Hypothesis as it occurred to us very vaguely in 1995 while we were working on another project.[1] The chapter reads like a travelogue because the idea developed in peripatetic fashion. With that idea in hand, we describe, in Chapter 2, surprising information about plants and animals, both the ones that share our dwellings and the ones we raise for food, showing a consistent pattern to the habitats where these organisms most likely evolved. Chapter 3 grapples with an idea that may cause a paradigm shift in many sciences: that humans are not simply a savanna species but one that used savannas in the context of the surrounding landscape, which included rock outcrops. We will show that only rock outcrops could provide the kinds of permanent environments that satisfied many of our resource needs. This leads us, in Chapter 4, to discuss the ecological properties of human dwellings, past and present. We will explore how these habitats select for exploitation by the plants and animals talked about in Chapters 2 and 3. Chapter 5 invokes E.O. Wilson's Biophilia Hypothesis to explain what we think is the spiritual significance of rock to many societies around the world, and Chapter 6 draws out the implications of that hypothesis for various aspects of our existence. In conclusion, Chapter 7 explores the idea of paradise courtesy of Wilson's hypothesis and ours.

Science is rarely a solo effort, and certainly every word of this book bears testimony to the talents and efforts of each of its five authors. Doug has been a research scientist and professor at the University of Guelph, Ontario, since 1975. He had previously studied the ecology of coastal tundra and began to work on rock lichens in 1976. When researching a large project dealing with a coastal lichen species, he began collaborating with Uta Matthes, who was finishing her PhD degree at Arizona State University. Uta joined the team in 1985 and helped to design some studies of cliff edge ecology during the summer of 1988. These studies led to the arrival of Peter Kelly, whose later discoveries of ancient cedar forests along the Niagara Escarpment garnered significant media attention across North America. John Gerrath's accidental discovery of microbial communities living inside the limestone rocks of the Escarpment resulted in his joining the team in 1992. As our work broadened into some areas of interest to theoretical ecologists and practitioners of restoration ecology, Jeremy Lundholm came on board, finishing his PhD in 2003.

Beyond these five, all of the other past members of the Cliff Ecology Research Group, centred at the University of Guelph, have participated in different ways to the formulation of the ideas in this book. Other faculty, including Brian Husband and Beren Robinson, have encouraged an evolutionary perspective to much of what we have written, and also insisted on the text being readable. Barrie Juniper of Oxford University, who has been following his own ideas about co-evolution between horses and fruit trees in Kazakhstan, was an invaluable ally and critic. David Evans, a molecular biologist at the University of Calgary with a very broad view of science and humanity, was especially helpful in correcting early mistakes about the evolution of the human-pathogen interaction. Ron Brooks of Guelph, who always thinks in terms of the big picture in ecology, provided unique insights. Tracy Sanderson of Cambridge University Press, who published the aforementioned book on cliff ecology, played a vital role, commenting on the work from the perspective of a publisher of anthropological material.

Many others, including Stephen Murphy, Larry Harder, Julie St. John, Hugues Massicotte, Darwyn Coxson, Brook Thiessen, and Melissa Purich, helped greatly by obtaining literature and simply discussing the project with us. Lastly, we thank Cameron Smith, Peter Taylor and the staff of Fitzhenry & Whiteside for their support and encouragement of this project. Funding for the work was provided by the Natural Sciences and Engineering Research Council of Canada.

So the book begins with a story about how we all – separately and together – fell upon the central idea of the Urban Cliff Hypothesis. At the start, it reads like a travelogue, because that's really what it was at the time. It involves trips into our backyards, as well as to places we'd never been before. But it all ended up as one story. It's a story about all of us – we the writers and you the readers. If true, it means that the Urban Cliff Revolution started a long time ago and will take us in the future to places both foreign and familiar.

---

1   We were writing *Cliff Ecology* (Larson et al. 2000a) at the time; then, later, we thought more about the work while doing fieldwork in various countries in 1997.

CHAPTER ONE

Travelogue
of an Idea

Virginia Larson was sitting around with her friends at her 10th birthday party, talking about their parents' jobs. One of them asked her about the work of her father's lab. She explained that the researchers there studied forests that grow on cliffs. "Why would they do that when there is so much forest growing on flat land that's easy to get to?" one friend responded. Virginia was speechless. Why *did* they go there? With all that flat, and safe, land to work on, why were they climbing all those cliffs?

The same question has been thrust on us, in slightly more subtle form, more than once in the nearly two decades since 1985 when the Cliff Ecology Research Group was formed. At one level, the answer is straightforward: cliffs are cool places that everyone admires. At another level, though, the answer is more complicated, that as scientists we've been trained to ask questions about cool things and figure out how they work. And we've come to discover some startling things about cliffs and their connections to people.

When we began to study cliffs, we often had to defend the work to colleagues who believed it was a bad idea to focus on unproductive land that was difficult or impossible to exploit commercially. We would always mount a spirited defence, claiming, for example, that the cliffs were distinct from what was below and behind them and that the ecological transition zones between cliff and surrounding landscape were interesting. But some found it curious if not perverse that we were studying places whose vertically projected area on aerial maps was zero.

Despite running into this attitude, between 1985 and the early 1990s we had published quite a few scientific papers on the ecology of cliffs. The most unusual and ultimately the most interesting feature of the cliffs we studied was that they supported ancient refuge forests with some of the oldest and slowest-growing trees in eastern North America.[2] As it turned out, the cliffs were unproductive refuge sites not only for trees but also for other plants, animals, and micro-organisms that took advantage of the general absence of human disturbance. In fact, the work ultimately led to a paradox: that land with no commercial value because it was so difficult for humans to exploit had become land of high ecological and conservation value because the organisms there were allowed to live in peace.[3]

From 1994 to 1996 we were working on a book called *Cliff Ecology*,[4] which was eventually published in 2000. The book, among other topics, covered the exploitation of cliffs, rock outcrops, and rock shelters at the bases of cliffs by archaic peoples. The general view has always been that humans were once cave dwellers or at least occasional cave users. At the very least, it has always been enticing to think of ourselves as crawling out of caves to achieve our current status as the most dominant species on the planet. In all this talk about cave men, however, few people had noted that most of the caves in question, rather than being openings in level ground, were located at the bases of rock outcrops and cliffs.

We also summarized work by British and Italian investigators who showed that man-made walls and stone buildings have many of the same macro- and microhabitats, and many of the same plants, as natural cliff faces. In the early stages of writing the book, however, we stopped well short of making wider claims about the importance of cliffs to people. In fact, in the first draft of *Cliff Ecology*, we ourselves accepted the idea that humans used cliffs no more than other landscape elements.

Cliffs as refuges for pigeons and rats

The year 1997 was a sabbatical year for Doug Larson, and freedom from teaching duties gave him time for extensive field trips in North America, Western Europe, and New Zealand. The purpose of these trips was to find out if other cliffs around the world supported relict ancient forests like the ones we had discovered in southern Ontario.

The travels began in late winter with Doug driving in a large arc from Ontario west to Michigan, Wisconsin, and Iowa, then travelling south to Tennessee and North Carolina. In Wisconsin, Doug was accompanied by Jeff Nekola from the University of Wisconsin at Green Bay. Jeff was a good one to hook up with since he knew Wisconsin and Iowa like the back of his hand. Doug and Jeff explored sections of west-facing escarpment along the edge of the Door Peninsula in Wisconsin. Doug was struck by the discovery that most of the plants and animals found on these cliffs were identical or similar to the ones growing on cliffs in Ontario, and that all these cliffs were refuges from human disturbance. The idea was beginning to form that cliffs worked the same way everywhere.

After these trips to the Door Peninsula, Doug set off to Iowa where he was to hook up with John and Jean Gerrath. About an hour's drive south of Steven's Point, Wisconsin, in an area that was once a large inland plateau, is a large sandstone tower known locally as Ship Rock. It measures about 20 m high, 60 m long, and 10–15 m wide at the base. No cities or towns of any size are in the vicinity. Doug had hoped that the face of this sandstone outcrop would be covered with stunted ancient woodland, like the sections of limestone escarpment in the Door Peninsula. Instead, Ship Rock supported a few white birch (*Betula papyrifera*) trees, several species of mosses that grow normally on acid rock, and some squirrels using small erosion pockets as food storage bins. Rock polypody (*Polypodium virginianum*) was growing all over the ledges of the main rock face, as well as on the large pieces of collapsed cap-rock at the base of the tower.

Doug was disappointed that there was no cedar forest on Ship Rock, but at least the other organisms were ones that "should have been there." But the big surprise was that without a city in sight, Ship Rock was enshrouded in a gray blanket of cooing pigeons; hundreds of them were roosting at the top of the outcrop. Every time Doug slipped on the icy rocks, and every time a dead branch snapped underfoot, the flock exploded into the air, settling back on the rock half a minute later. The contrast with the silence of the surrounding forest was intense, and each of the flurries of pigeon activity made Doug feel uncomfortable.

The pigeon experience was locked away in short-term memory, and the road trip continued on to the Gerraths and Cedar Falls, Iowa. John Gerrath was going to help him hunt for ancient forests on cliffs in regions of Iowa with surface rocks similar to Wisconsin and Ontario. They had time prior to these field trips, however, to visit the University of Northern Iowa's library to consult the literature on birds. Doug found accounts of several pigeon and dove species that exist as endemics[5] on rock outcrops, cliffs, and coastal mountains in the Indian subcontinent. He also read that among all of these species, only the common pigeon – more correctly called the rock dove (*Columba livia*) – showed explosive population growth and range expansion when it made contact with humans.[6] Why similar, related species living in the same habitat have remained in small isolated populations was not known.

One thing was clear, however. The pigeon, a bird that most people think of as an obligatory city-dweller, actually has an evolutionary past that precedes the Holocene (the last 10,000 years). Its evolutionary arrival occurred at a time when there were no cities, no statues, and no people; it had a natural habitat that preceded all of these. We are now most familiar with it in an urban landscape (Figure 2), but the pigeon at one time was a wild bird like any other. Only recently in our evolution has it come to adopt us.

Figure 2.
The frontispiece of Simms (1979) emphasizing the homology
between the natural and built environments that support pigeons.
Original artwork: Peter Knock. Used with permission.

This information provided a foundation for interpreting Ship Rock. It probably reflects the habitat setting within which pigeons evolved in India: isolated rock outcrops surrounded by productive forest rich in the seeds of nutritious plants. Ship Rock might, therefore, be normal and the pigeons of cities the aberration. Pigeons in cities might simply be exploiting a habitat that provides for their needs: food and suitable breeding or roosting sites. The food is normally supplied by plants, and the breeding and roosting sites must be flat ledges of hard material elevated well above the ground.[7] Brick, concrete, or glass buildings will do, as will statues. The pigeons just want high rocks. *Any* high rocks.

The other members of the research group greeted Doug's travel reports and new insights with excitement. They pointed out that cities have been described as concrete canyons for at least a hundred years, and added that any granivorous bird would do well to follow humans to their cities because of the huge amounts of stored and wasted food we create.

The next day's field trip continued the search for cliffs and ancient forests. Driving from one field location to another can be a little boring in Iowa. Even though steep river canyons are frequent in the northeastern part of the state, the land between the canyons is nearly flat. The initial land surveyors made a point of carving out counties with a regular grid of perfectly straight roads. John and Doug, attempting to amuse themselves until reaching the next field site, began to talk about the pigeons again.

It occurred to them that the rock dove might easily be the most common bird in the world as a consequence of people having built sites that function perfectly as cliff habitat. They then jumped to the question: "What other organisms are now very populous and associated with humans? Where did they come from?" The draft copy of *Cliff Ecology* already contained the observation that many plant species found on stone walls were once cliff endemics but had switched to man-made habitats. While many of them were weedy species, not all of them were.[8] Certainly, we humans took them all for granted, or worse, we disparaged their presence. The rock polypody fern (*Polypodium vulgare*) and the wall rue (*Asplenium ruta-muraria*) came immediately to mind. At this point in the conversation, another question came up: "So where do rats and mice come from?" John suggested jokingly that this vague idea should be called the Rat Hypothesis. At that very moment, both Doug and John realized that this might not be a joke. Even if rats were not involved at all, the *idea* that rats (*Rattus rattus* and *Rattus norvegicus*) or any other species might exploit man-made environments for the simple reason that they resemble the places where they originally evolved was taking root. The roots were far deeper and far more fibrous than we could imagine.

On to Appalachia!

The Rat Hypothesis was not explored further in Iowa, but the next stop in the giant circle tour of eastern North America – Boone, North Carolina – proved useful. Appalachian State University is situated in an exquisite small valley high in the Blue Ridge Mountains. Increasingly, Boone's reputation as a mountain paradise town is resulting in an erosion of its idyllic character, with mini-malls rapidly replacing quaint shops. This trend disturbs Gary Walker, who lives in Boone and works in the biology department at Appalachian State. Gary and Doug had been in professional contact for some time because Gary had studied the ecological genetics of small, scattered populations of eastern white cedar (*Thuja occidentalis*) on cliffs in Tennessee. Gary didn't know it when he was completing his study in 1987, but his work on the genetics of this species, and more importantly on the other species associated with white cedar on the cliffs, formed a vital link with the ideas being developed by the Cliff Ecology Research Group.

Douglas Mielke, an expert on the physiology of the house mouse, was the chair of biology at Appalachian State. On the basis of the sudden interest that we had in rodents, Doug Larson asked him for his reaction to the central idea of the Rat Hypothesis. Mielke had never considered the natural history and evolution of mice prior to their contact with humans. He suggested that the library might provide evidence on current and past records of wild rodent populations around the world. He did caution, however, that most work on the house mouse (*Mus musculus*), black rat (*Rattus rattus*), and Norway rat (*Rattus norvegicus*) followed the introduction of these species into Europe in the Middle Ages. Indeed, the first attempts to find information about the natural history of these organisms before the time of Marco Polo were unsuccessful. Far more people were working on the genetics of rats and mice than on their natural history: interest in where they came from was minuscule compared with interest in how to use them or how to get rid of them.

Larson contacted Bill Jackson of Bowling Green State University in Kentucky, a world authority on rats, and sketched the outlines of the Rat Hypothesis. Bill was skeptical. He suggested that (1) there was little literature on the original habitats of rats and mice prior to contact with Europeans in the 13th century, (2) using molecular markers to reconstruct the sites of origin would be possible but very expensive and would likely offer no commercial product to anyone, and (3) it was very likely that woodlands and forests were the habitats of origin of these species anyway.

Several more days of digging through the library at Appalachian State yielded some important papers. One of them[9] described wild populations of black rats and house mice living on talus slopes[10] and cliffs in mountainous coastal areas of Libya. Incidentally, these were the same habitats where *Felis silvestris* was found, the European wildcat from which domestic cats have evolved. The rodents lived far from human settlements, and researchers concluded that these were natural populations breeding under their own control. Older literature[11] suggested that feral black rats would revert to wild habitats that included talus slopes and the bases of cliffs in Pennsylvania, and yet another report was found[12] stating that many small rodents were extremely efficient exploiters of rocky terrain, and that many of the rarer species of rat-like rodent in eastern North America were only found in areas with rock outcrops or escarpments.

Later, when these results were communicated to Bill Jackson, he softened his initial resistance to the Rat Hypothesis, admitting the possibility that rodents now associated with humans were once part-time or full-time colonists of cliffs, rock outcrops, or other low-elevation rocky land-scapes. He pointed out that rocky sites usually rise slightly above the surrounding forested landscape and offer access to a rich food supply, refuge from predators, sites for food storage, and excellent breeding dens all at the same time. In later correspondence, Bill agreed that the idea might be correct after all, but offered that it would still be frightfully difficult to prove. The Rat Hypothesis would rest for a while.

European vertical wilderness

With summer and fall came field trips to Germany, France, and the United Kingdom. The goal of these trips, like the earlier ones, was to compare cliffs in other temperate locations with those in southern Canada. The literature generally agreed that virtually all of Western Europe had been cleared of its original vegetation during the past 3000[13,14] years. While planning the trip in 1995, Doug had corresponded with Daniel Barthélémy, a botanist in Montpellier, France, who had indicated that there were indeed wonderful tall cliffs in the southwest of France, but warned, "I think that there is no equivalent here of your fantastic forest growing on the Niagara Escarpment." Daniel was pleased to write two years later that "the age of the trees [on the cliffs] is really quite impressive and I am really happy to know that I was totally wrong." Besides ancient forests, Doug found that many of the species and most of the genera of plants found on cliffs in Europe were similar to the ones seen in North America. There were, however, some fascinating differences.

In Germany, Doug learned that a surprising number of colourful, horticulturally important plant taxa were in natural settings restricted to rock outcrops and cliffs. Among them were several species of carnation (*Dianthus*), the cliff speedwell (*Veronica fruticans*), the German catchfly (*Lychnis viscaria*), and the edible caper (*Capparis spinosa*). Cliffs were also the original habitat of several species normally considered weeds, such as bladder campion (*Silene vulgaris*), Kenilworth-ivy (*Cymbalaria muralis*), and wall rue (*Asplenium ruta-muraria*). Doug's European hosts were most gracious in pointing out these plants in their natural rocky habitats, but they were also a little amused by Doug's seemingly excessive interest, given the fact that these plants were so universally abundant in habitats created by people.

While entertaining Doug at his house, Wilhelm Barthlott of the Botanisches Institut in Bonn listened intently to his visitor's description of the rock polypody (*Polypodium virginianum* or *P. vulgare*) as a wonderful desiccation tolerant cliff plant quite uncommon in Canada. He then ever so gently, and perhaps even with a slight sense of discomfort, opened a sliding glass door to reveal the same species growing in the cracks of his patio.

An encounter with a goat

A paradox was developing. From the perspective of *Cliff Ecology*, cliffs were refuge habitats occupied by tough, long-lived species that lacked competitive ability. We had concluded in the book draft that cliffs were one of the least disturbed terrestrial habitats on earth and had urged that conservation efforts be made to protect them. Yet, below the sparse canopy of ancient juniper (*Juniperus*), oak (*Quercus*), and yew (*Taxus*) was an understory containing a combination of horticulturally important species and weedy species. The pigeons on Wisconsin's Ship Rock and the rats of north Libya also seemed out of place. How could perpetually undisturbed habitats select so strongly for organisms that had weedy opportunistic life histories? Even more questions were raised by an encounter with a goat.

On a hot July afternoon, Stefan Porembski was leading Doug on a field trip along the Nahe River valley near Bonn. It had been a five- or six-hour hike, and Doug was tired; not so Stefan, an expert on the biodiversity of isolated granite outcrops called inselbergs, who was fit from walking vast distances up and down rocky slopes. Toward the end of the day, the two encountered a cliff of volcanic rock about 10 metres high that supported some old-looking junipers on a ledge about 2 metres over the crest. As Doug was sitting down at the cliff edge to assemble equipment for aging the trees, a visitor came into view. It appeared to be a wandering domestic goat (*Capra hircus*, Figure 3). Goats do not show a huge variety of facial expressions. Fear, elation, and utter boredom, for all we know, are indicated by the same blank stare. One should never play poker with a goat. The goat walked to the edge of the cliff where Doug was sitting. Because of its domestic status, nobody was particularly wary of it, but it was fun to imagine it as a Chamois foraging on low-elevation cliff edges in the Alps. For less than a minute the goat nibbled at the sparse cliff-edge vegetation. As Doug extracted his camera for some quick photos, the goat moved slowly, like the minute hand of a clock, to a position that placed Doug between the cliff edge and itself. At this point, the goat's head went down, and it started moving toward Doug . . . and this book was very nearly never written.

Figure 3.
This free-ranging goat (*Capra hircus*) was foraging along a cliff edge in the Nahe River valley of western Germany. Photo: D. Larson.

That evening over wine and food, the episode seemed more comical than informative. But it did raise the question, where do goats come from? And, for that matter, where did other animals come from that have been domesticated? Doug would ruminate on these questions until the trip to Great Britain.

Whence cabbages?

By September, the deck was cleared for the next trip. This one involved working with botanist Alan Charlton of Manchester University. He and his wife, Anne, had offered their house as a base camp for field trips in north-central United Kingdom.

Wales is a country with a strong connection to rock. Sedimentary limestone, volcanic rocks, and basaltic pillars are all present in different places. Mount Snowdon dominates the sky and rocky coasts are ubiquitous. Doug had selected several limestone cliffs in the northern part of the country for his study, among them Llangollen and the Great Orm at Llandudno. Both of these turned out to support stunted and slow-growing forests, but the Great Orm (Figure 4) was particularly interesting. It is circumscribed by a roadway system that allows visitors to access the top of the limestone pediment by car. Human pedestrians are discouraged, but sheep walk on the roadways and graze freely on the sloping ground.

Stunted yews occur on the limestone cliffs facing out to the sea, and the Hart's tongue fern (*Phyllitis scolopendrium*) grows on large blocks of collapsed rock at the base of the huge cliffs.

At one point, while hiking slowly along the base of the cliffs, Doug spotted a herbaceous plant growing from a tiny rock ledge.

Figure 4.
The Great Orm, Wales. The stark-looking limestone cliffs support populations of ancient yew trees as well as populations of cabbage (*Brassica oleracea*) growing in their natural habitat (see Figure 5).
Photo: D. Larson.

The leaves, with prominent veins and a faintly grayish-blue-green waxy appearance, looked oddly familiar. It looked like cabbage. Alan confirmed that the plant was cliff cabbage or sea cabbage and was somehow related to the horticultural variety. Seeing Doug's sudden interest, Alan added that lots of plants we eat, such as lettuce, beets, and onions, have relatives that occur spontaneously on cliffs or rocky talus slopes.

Background library work on Doug's return to Manchester revealed an even closer connection between cliffs and the human food system than Alan had suggested. The cliff cabbage was actually *Brassica oleracea* (Figure 5) and was exactly the same species as cabbage, brussels sprouts, broccoli, and cauliflower.[15] All of these vegetables were simply varieties of cabbage, and all were descendants of a plant that grew spontaneously on coastal cliffs from the Mediterranean to the North Atlantic. Other human food species, for example, beets (*Beta vulgaris*), were also native to rocky shorelines in the same areas. The site at Llandudno where Doug had seen the cliff cabbage could hardly have supplied enough food for foraging humans, but the plants could be transplanted to other locations and could be cultivated and selected to grow rapidly in situations where competition and other features of the plant and its environment were controlled by people. In other words, this plant could be domesticated.

So here was a plant that was a strict cliff endemic, but was also opportunistic enough, productive enough, and weedy enough to tolerate exploitation by people. The cabbage had taught us a tasty lesson.

Figure 5.
Cabbage (*Brassica oleracea*) growing in its natural cliff habitat. The rocky surface is part of the Great Orm in northern Wales. Photo: D. Larson.

Markland Grips

A week later Doug travelled toward Sheffield to visit a site he had learned about from a famous paper published by Jackson and Sheldon in 1949, the year he was born. The paper described interesting vegetation patterns across limestone cliffs on the Chatsworth Estate. The cliffs turned out to support truly ancient woodland, but again, that story is an aside. Along with large and magnificently deformed yew trees and ancient lime trees (*Tilia cordata*), the Chatsworth cliff faces supported a wide variety of herbaceous plants, including many that Doug had seen in Germany: wall rue (*Asplenium ruta-muraria*), cliff lettuce (*Lactuca muralis*), nightshade (*Solanum dulcamara*), and dandelion (*Taraxacum officinale*). The talus slopes also had interesting and familiar plants, including elderberry (*Sambucus nigra*), blackberry (*Rubus fruticosus*), hawthorn (*Crataegus monogyna*), speedwell (*Veronica sp.*), and burdock (*Arctium vulgare*). Not only were these plants becoming a familiar sight, but it was also increasingly clear, in Europe at least, that cliff and talus species were *the same as*, or *similar to*, species exploited in agriculture (i.e., crops) or otherwise associated with humans (e.g., weeds).

As the field trip to Markland Grips was wrapping up, Doug consulted the ordinance survey map on the topography of the site he had just visited. The map listed, about two kilometres to the east, a village called Creswell and within the village a rocky canyon with the label Creswell Crags. Doug and Alan could have walked there in 20 minutes. Doug thought that the name was vaguely familiar, but it had started to rain and there was neither the time nor the inclination to pay a visit to the site. That was a mistake.

Unbeknownst to Doug and Alan, the site they had decided not to visit was one of the best known and earliest human habitation sites in western Europe. Creswell Crags was used continuously by *Homo neandertalensis* from 80,000 to 40,000 years ago, and later by *H. sapiens* 20,000 to 8000 years ago. During these periods, the human habitations would have been surrounded by all of the plants and animals that were the normal residents of rocky environments. A serendipitous bond could have formed among these organisms, a bond that might still be reflected in the kinds of habitats, flora, and fauna that surround human settlements around the world today. The development of the Urban Cliff Hypothesis would have taken a leap forward if Doug and Alan had taken that 20-minute walk. Instead, the hypothesis simply fermented along with the cliff cabbages, the rats, and the pigeons.

The birth of the Urban Cliff Hypothesis

We were all feeling a little disquieted by the Rat Hypothesis as time passed. Further conversations with Bill Jackson and others had suggested that the precise habitats of origin of rats and mice might be impossible to determine. In contrast, we had made significant progress with cabbages and pigeons but did not want to restrict the hypothesis by renaming it the Pigeon Hypothesis or, worse still, the Cabbage Hypothesis. So the group assembled one afternoon, trying to come up with a new label by visualizing what the Rat Hypothesis was all about. Some new names were suggested: Cliff Biota Adoption Hypothesis, Cliff Habitat Replication Hypothesis, Cliff Replication Hypothesis, Cliff Biota Naturalization Hypothesis. None of these worked. Then Pete Kelly said, "What about Urban Cliff Hypothesis, since the core of the idea is that humans continue to create environments that represent their original rock shelters?" The Urban Cliff Hypothesis was born.

Timing the release of a new idea is always difficult. In 1997, we felt that if we didn't include the hypothesis in *Cliff Ecology*, we'd miss the chance to make readers aware of the possible ecological connection we had discovered. At the same time, we knew we had not done the necessary homework; we needed to research the background literature that the hypothesis rested on completely and rigorously before presenting it to the public. Hence, the book offered up the hypothesis as an interesting but untested idea. By December 2001, however, many readers had corresponded with us regarding *Cliff Ecology* in general and the Urban Cliff Hypothesis in particular. Most felt that it was an attractive if controversial idea. After discussions with many different people, we finally decided to evaluate the hypothesis ourselves. This book is the result of that evaluation.

In its briefest possible form, the Urban Cliff Hypothesis states that cliffs, talus, and other rock outcrops represent a marginal, unproductive refuge habitat that gave rise to or represented an opportunity for a group of species that are now central and highly productive: humans and their commensals and mutualists.[16] The rest of this book assembles evidence that relates to the hypothesis. This evidence falls into four broad categories, which will be presented in turn: (1) the array of plant and animal species that have been our companions through history and their habitats of origin; (2) the conventional view of human evolution during the Pleistocene and the alternate view provided by the Urban Cliff Hypothesis;

(3) the ecological properties of human dwellings in the context of the plants and animals discussed earlier; and (4) the biological basis of the attraction that most people have to cliffs and rock. We feel that the hypothesis is revolutionary in that it binds together an enormous amount of the human experience in one story.

2   An article by us reviewing the idea of cliffs as refuge habitats was published in *American Scientist* (Larson et al.1999).
3   An article by Stahle and Chaney (1994) came to a similar conclusion for post oak forests in the central United States.
4   Larson et al. (2000a).
5   An endemic is a species that is native to a particular region, and geographically restricted to it.
6   Darwin (1859).
7   Johnston (1992). Richard Johnston is the residing expert on pigeons. He has also co-authored the definitive book on the biology of feral pigeons (Johnston and Janiga 1995).
8   Pysek et al. (1995). This book, *Plant Invasions,* is one of many volumes in recent years that focus attention on the global problem of native biotas being replaced by weedy ones.
9   Ranck (1968). In addition to rats and mice, Ranck also reported gerbils in habitats adjacent to the rocky escarpments along the north coast of Libya.
10 Talus slopes, also called scree slopes, are accumulations of loose rock debris found at the base of cliffs.
11 Audubon (1989). This is the reprinted edition of the famous volume *The Viviparous Quadrupeds of North America,* originally published in 1854.
12 Feldhammer et al. (1984). Like many biological inventories, this publication is difficult to get without contacting the state government office involved.
13 Peterken (1996).
14 Spencer and Kirby (1992).
15 Zohary and Hopf (1993). This volume and Sauer (1993) both offer up to date accounts of the regions, and in some cases the sites and habitat types, of origin for many crop plants.
16 Commensalism and mutualism are two different ways in which pairs of associated organisms can interact. In commensalism, one participant benefits from the relationship while the other derives neither benefit nor harm. In mutualism, both participants benefit. These terms are explained in more detail in Chapter 2.

CHAPTER TWO

Tulips, Pigeons, and People: Origins of the Human Ecosystem

Here is a challenge. Put this book down. Go outside and record the names of all the plants or animals that you see. Actually, start with the organisms that are right where you are, and those you pass by on your way out. In our lab, house spiders have set up housekeeping right beside us under the desks, and house mice navigate the walls, so many of them that we cannot leave our chocolate bars unsecured overnight. There are also centipedes, silverfish, pillbugs, and German cockroaches in the cracks of the walls. Outside the building, the first woody plant we see is a tree-of-heaven, followed closely by a Virginia creeper and a wild grape. Wild lettuce and common groundsel are growing in the cracks of the pavement. Pigeons, house sparrows, and starlings are circling overhead trying to stay clear of the kestrel that has made its nest on a small ledge of the physics building nearby.

Go to your local produce market or grocery store and you'll find a wide variety of fruits and vegetables, which in northern latitudes will include apples, pears, cherries, melons, peppers, beans, potatoes, and tomatoes (Figure 6). Herbs such as parsley, mint, oregano, and dill may also be found. You'll also see a wide variety of flours made of wheat, corn,

Figure 6. A photograph taken at the farmer's market in Guelph, Ontario. Potatoes, onions, rhubarb, carrots, and peppers are just some of the crop species that are derived from ancestors that lived in cliff, talus slope, and other rocky outcrop habitats. Photo: D. Larson.

oats, rye, and rice. In the meat department you'll find beef, pork, lamb, and a variety of fowl.

It is admittedly difficult to imagine, but all of these species were once part of nature without people. They are all the direct descendants of wild organisms, and in fairness to them, they may still "think" they are wild. It is only we who consider them cosmopolitan species with an affinity for human habitats.

Where have these organisms come from? This question has long fascinated biologists. One of the first to investigate the origin of cultivated plants was de Candolle, followed by Vavilov in the early part of the 20th century and, more recently, Baker, Simmons, and co-workers.[17] The conventional view holds that if these species weren't somehow "created" by humans, then they originated around the Mediterranean Sea and east to India and China, first becoming abundant in Mesopotamia. As we will show here, that is barely the beginning of the story.

The biodiversity associated with people, then and now

The media draw our attention to the global loss of biodiversity, but we are actually surrounded by a surprising amount of biodiversity, even in cities.[18] Of course, most of us hate, dislike, or are concerned about this biodiversity for a variety of reasons.[19] Its has been observed for a long time that indigenous plants and animals are being rapidly displaced by an assemblage of weedy, undesirable species that are similar around the world.[20] These organisms are tolerant of humans, or even dependent on them. As Peter Ward believes,[21] they are likely to become the biological base for the future evolution of many species.

The obvious but rarely asked question is where all the biodiversity came from. Where were house mice, house sparrows, house centipedes, barn owls, barn swallows, wall-flowers, wall lizards, and bed bugs *before* there were houses, barns, walls, and beds? What is the original habitat of the dozens of weed species that we spend millions of dollars each year trying to eradicate: dandelions, plantain, catchflies, thistles, docks, pigweeds, lamb's-quarters, and mustards? We believe that it is necessary to understand the ecological history of the organisms that are part of this global invasion in order to decide whether, and how, to stop it. This decision of course involves value judgements based on the Western cultural affection for "nature in the raw." To many non-Western cultures, nature is still the enemy, to be destroyed and replaced with exploitable organisms as rapidly as possible. However, in this book we will put aside the question of whether and how to prevent the biotic homogenization of the world. We will focus entirely on the origin of the plants and animals that have been and are now associated with humans.

We begin this chapter with the compilation of a list from the prehistorical record of organisms whose remains tend to be associated with human remains at human dwelling sites. We will then identify plant and animal species that are closely associated with modern humans. For all these species, we will try to identify the habitat of origin, or if such information is lacking, the habitat where these species (or their wild relatives) are found today.

Prehistoric times

Animal and plant remains found in fossil campsites may suggest that these critters were eaten by people, but they do not accurately reflect the full composition of our ancestors' diet,[22] because animal parts (bones, teeth) are more easily preserved than plant parts. And even the plant parts that remain (seeds and pollen) only indicate that the plants were there. Their presence in a site does not prove that the species of plant was gathered and eaten.

The remains of vertebrate animals are almost always recovered at sites that were used by early humans (*Australopithecus* and *Homo*). A small proportion of these[23] are carnivores such as cave lions, tigers, civet cats, hyenas, foxes, wolves, and dogs that very suddenly replaced wolves at many sites.[24] Omnivores, including cave bears, brown bears, langurs, gibbons, and other primates, are commonly found, as are bats and predatory birds such as peregrine falcons. The usual interpretation of these findings is that the same sites used as shelters by humans were also selected as dens by these animals,[25] although some argue that the predators may also have preyed on humans. There are few suggestions in the literature that humans ate any of these species.

Sites used by early humans also commonly contain the skeletal remains of herbivores and granivores. Many of these were most likely consumed or otherwise used by humans. The list of larger species from periods of climatic cooling includes reindeer, mammoth, bison, and rhinoceros. Remains from warmer periods are dominated by cattle, horses, pigs and boars, sheep, goats, and in South America, llamas.[26] Small herbivorous or granivorous mammals such as rabbits, guinea pigs, porcupines, rats, and mice have also been frequently found,[27] and there are occasional reports of granivorous birds such as pigeons.

Animal remains are common at human occupation sites, but the actual number of species is much smaller than the total number of animals in the entire regional fauna that existed at the time. Aurochs or cattle, horses, pigs, deer, mountain goats, and ibex[28] make up the vast majority of the animals associated with human remains; Potts (1984), for example, reports that 84% of remains in Africa are from cattle, horses, and pigs. The same species are found consistently from continent to continent wherever human remains have been discovered, and at some sites such as Spirit Cave and Banyan Valley Cave in Thailand, the small array of granivores reported was constant from 14,000 years ago until 2000 years ago when cave use ceased.

Much information on the plant species that coexisted with early humans comes from pollen analysis.[29] Among the woody plants identified are junipers, pines, elms, oaks, willows, buckthorns, alders, olives, and almonds. In addition, pollen, spore, or seed material of chamomile, thalictrum, crowberry, rice, emmer wheat, einkorn wheat, barley, other grasses, lentil, pea, composites, potato, sweet potato, common bean, teosinte, maize, pepper, betel, spikemoss (*Selaginella*), and rock polypody (*Polypodium*) have been identified in cave deposits associated with human settlements.[30] Pollens from a wide spectrum of plants native to the savanna or upland habitats in Africa, Asia, Europe, and South America have also been reported, but in much smaller amounts than for the plants listed above.

As with the animal remains, this list of plants is clearly a small and non-random subset of the large number of species that occurred in the areas surrounding the human occupation sites.[31] This shows that a particular group of species[32] was being selected by humans, who perhaps were foraging broadly from their home bases to collect wild plants, returning them to their dwellings for processing.

Present times

Ecologists define interactions between pairs of organisms by distinguishing whether the organisms involved benefit from the interaction, are harmed by the interaction, or derive neither benefit nor harm. The relationship that humans have with many of their associated species can be classified as either *commensal* or *mutualist*. *Commensalism* is an interaction in which one participant derives a large benefit while the other derives no measurable benefit or harm. Examples of animal and plant species that are commensal with humans in most parts of the world are listed in part

(a) of Table 1. Vertebrates include, by order of abundance, rats, house mice, pigeons, starlings, cliff swallows, house sparrows, and barn owls.[33] Most of these are among the most cosmopolitan and abundant species within each of their respective genera. Invertebrates commensal with humans include the bedbugs, and plants commensal with humans include the several hundred species considered weeds, although the latter can also be viewed as competitors to our crops (Figure 7). Members of the sunflower and mustard families are especially well represented among these.

Figure 7.
Dandelion (*Taraxacum officinale*) growing on a limestone cliff in southern Ontario. Native to cliffs and talus slopes in Europe and western Asia, dandelion grows as a weed in cities in Europe and North America but also colonizes natural cliff faces throughout its range. Photo: D. Larson.

Ecologists define interaction between two organisms that results in a benefit for both participants as *mutualism*. Some familiar examples of this are listed in part (b) of Table 1. For the label *mutualism* to apply, neither the human nor the companion species may kill and consume the other (if one participant is harmed to the benefit of the other, the relationship is *predation*). Note that these definitions are based on what happens to the individual organisms involved. As a consequence, it is possible for some individuals of a species to be mutualistic with humans, while other individuals of the same species are exploited by humans in a predatory relationship. Examples of species whose relationship with humans can be based on either mutualism or predation include cattle, horses, asses, sheep, and goats.

TABLE 1

Animal and plant species are currently associated with humans in one of three ways: (a) commensal with humans (benefit from the relationship with humans while humans are little affected); (b) mutualistic with humans (the organisms and humans both benefit from the relationship); and (c) exploited by humans in agriculture. For each species we have compiled information from the literature about their place and habitat of origin. We list the place and habitat where the species is currently found growing in the wild, where close relatives of the species are currently found growing wild, or where fossils of the species have been retrieved. Organisms are identified by both their common English name and their scientific name and listed in the approximate order of their global importance.

COMMON NAME	SCIENTIFIC NAME	PLACE AND HABITAT OF ORIGIN

(a) Commensal organisms
Animals

COMMON NAME	SCIENTIFIC NAME	PLACE AND HABITAT OF ORIGIN
Black rat	*Rattus rattus*	Original habitat not known, but fossils found in caves associated with *Homo erectus* in eastern Asia (Aigner 1978a). Unidentified species of rat also found associated with *Australopithecus boisei* (Leakey 1960). Became commensal after the first permanent settlements were established (Budiansky 1992). Ranck (1968) reports wild *Rattus* from limestone escarpments and talus slopes in Libya
Norway rat	*Rattus norvegicus*	Migrated from Asia in the early 18th century (Budiansky 1992). Habitat unknown
House mouse	*Mus musculus*	Fossils dated to >600,000 years B.P. found in caves with *Homo erectus* (Aigner 1978b)
Pigeon	*Columba livia*	Once endemic to rock outcrops and cliffs in central Asia (Johnston 1992). Fossils dated to 13,000 B.P. also reported from sea caves in U.K. (Smith 1992)
Starling	*Sturnus vulgaris*	Nests in cavities in cliffs and burrows (Cabe 1993)
Cliff swallow	*Hirundo pyrrhonota*	Makes nests on vertical cliff faces in the mountains of western North America (Brown and Brown 1995)
House sparrow	*Passer domesticus*	Common in grasslands enriched by rock outcrops (Lowther and Cink 1992)
Barn owl	*Tyto alba*	Endemic to rock outcrops and cliffs in Africa and Asia. Largely a predator on rodents (Vrba et al. 1995)
Bedbug	*Cimex lectularius*	Rocky habitats in Africa and Asia. Originally ectoparasites of pigeons and bats that lived in caves (Dolling 1991)

Plants

COMMON NAME	SCIENTIFIC NAME	PLACE AND HABITAT OF ORIGIN
Dandelion	*Taraxacum officinale*	Talus slopes, rocky waste places in low mountains (Stace 1999)
Plantain	*Plantago* species	Rocky ground, cliffs, waste places (Stace 1999)
Hawkweed	*Hieracium* species	Escarpments, cliffs, steep slopes, waste places (Stace 1999)
Wall rue	*Asplenium ruta-muraria*	Rock outcrops, escarpments, rock walls (Ellenberg 1988)
Rock polypody	*Polypodium vulgare*	Rock outcrops, escarpments, rock walls, tree branches (Ellenberg 1988)
Kenilworth-ivy	*Cymbalaria muralis*	Rock outcrops, escarpments (Ellenberg 1988)

(b) Mutualistic organisms
Animals

Horse	*Equus caballus*	See below under (c)
Ass	*Equus asinus*	See below under (c)
Cattle	*Bos taurus*	See below under (c)
Sheep	*Ovis orientalis*	See below under (c)
Goat	*Capra hircus*	See below under (c)
Dog	*Canis familiaris*	Fossils found in close association with those of wolves (Smith 1992); transition from wolf to dog is almost instantaneous. First vertebrate consistently associated with human encampments (Morey 1994). Central part of Andean cultures by 8000 years B.P. (Wing 1977)
Cat	*Felis domesticus*	Escarpments and rock outcrops in Egypt, Palestine, and the Mediterranean (Kurtén 1965b)
Elephant	*Elephas maximus*	Open forest and savanna in east Asia (Nowak and Paradiso 1983)
Cockroach	*Blattella germanica*	Talus slopes in Africa; detritus feeder (Rust et al. 1995)

Plants

Hemp	*Cannabis sativa*	Human dump and manure sites in early encampments. Followed humans in the early phases of their evolution (Löve 1992)
Tulip	*Tulipa* species	Rocky screes in eastern Turkey; pockets between limestone crags; deep soil among limestone rocks (Pavord 1999)
Geranium	*Geranium* species	Rocky screes in eastern Turkey (Pavord 1999); cliffs and rocky slopes (Stace 1999)
Geranium	*Pelargonium* species	Rock crevices in southwest Africa (Wilson 1946); cliffs and rocky slopes (Stace 1999)
Forsythia	*Forsythia* species	Rocky habitats (Wu and Raven 1996)
Peony	*Peonia* species	Rocky habitats (Fairchild 1919)
Petunia	*Petunia* species	Exposed spaces between rocks on steep canyon walls that are formed by rivers cutting through rock strata in South America (Ando and Hashimoto 1998, Sink 1984)

(c) Organisms exploited in agriculture
Animals

Cattle	*Bos taurus*	Open sloping grasslands in mountains (Budianski 1992, Diamond 1999)
Horse	*Equus caballus*	Open sloping grasslands and fruit forests in mountains (Budianski 1992, Diamond 1999)
Ass	*Equus asinus*	Open, highly productive grasslands in mountains (Budianski 1992, Diamond 1999)
Sheep	*Ovis orientalis*	Mountain slopes in western Asia. Used either wild or domesticated since 15,000 years B.P. (Bogucki 1996)
Goat	*Capra hircus*	Steep slopes and cliff ledges at higher elevations in western Asia (Carr 1977). Used since 15,000 years B.P. (Bogucki 1996)
Pig	*Sus scrofa*	Fossils recovered from caves in eastern Europe and western Asia (Reed 1977a,b, Mithen 1990)
Guinea pig	*Cavia porcellus*	Steep slopes of the Andes (MacNeish 1977). Used (but not domesticated) by 9000 years B.P. (Wing 1977)
Fowl	*Gallus gallus*	Tropical forest of east Asia and India (Stevens 1991)

Plants

Emmer wheat	*Triticum dicoccoides*	Hard limestone bedrock, limestone slopes, talus (Zohary and Hopf 1993); mountain ravines, fissures in rocks rich in lime, humus, and manure (1992). Fossils dated to 15,000 years B.P. found in Greece (Bogucki 1996)
Durum wheat	*Triticum durum*	Limestone slopes and outcrops, associated with encampments (Bogucki 1996)
Einkorn wheat	*Triticum monococcum*	Steppe and open oak woodland (Zohary and Hopf 1993)
Bread wheat	*Triticum aestivum*	Steppe, open areas (Ryan and Pitman 1998, Zohary and Hopf 1993)
Rice	*Oryza sativa*	Steeply sloping land well supplied with water; warm wet marshes (Zohary and Hopf 1993); fossilized grains dating to 4700 years B.P. found in caves in Thailand (Gorman 1977)
Barley	*Hordeum vulgare*	Steppe, open woodlands, grassland (Zohary and Hopf 1993); wild barley reported from wadis between river channels and the bases of cliffs in Egypt (El Hadidi et al. 1986)
Corn	*Zea mays*	Open oak woodland, limestone outcrops and slopes (Sauer 1993)
Teosinte	*Zea mays* ssp. *parviglumis*	Open rocky limestone slopes (Beadle 1977)
Rye	*Secale cereale*	Plateaus on mountains, slopes, escarpments (Zohary and Hopf 1993); escarpments in the central Mediterranean basin (Löve 1992)
Oats	*Avena sativa*	Slopes in western Asia (Zohary and Hopf 1993)
Millet	*Panicum miliaceum*	Desert margin and savanna in northern Africa (Sauer 1993)
Sorghum	*Sorghum bicolor*	Desert margin, savanna (Sauer 1993)
Potato	*Solanum tuberosum*	Heath, slopes of Lake Titicaca, Peru and Bolivia (Sauer 1993) Wild *Solanum* reported from cliffs in coastal Peru (de Candolle 1964)
Sweet potato	*Ipomoea batatus*	Unknown (Sauer 1993)
Cassava	*Manihot esculenta*	Arid, seasonally dry grasslands (Sauer 1993)
Asparagus	*Asparagus officinalis*	Rock outcrops and talus slopes in Asia and the Mediterranean (Löve 1992)
Bean	*Phaseolus vulgaris*	Arroyo banks and rocky limestone slopes (Sauer 1993)
Broad bean	*Vicia faba*	Fully exposed, rocky places in upland areas (Zohary and Hopf 1993)
Pea	*Pisum sativum*	Steppe and open woodland (Zohary and Hopf 1993). Fossils associated with cave sites around the margins of the Mediterranean dated to 15,000 years B.P. (Bogucki 1996)
Chickpea	*Cicer arietinum*	Limestone bedrock, scree, and rock outcrops in Turkey (Zohary and Hopf 1993)
Lentil	*Lens culinaris*	Shallow stony soil, stone patches, heaps, barren rocky slopes (Zohary and Hopf 1993, Sauer 1993). Fossils associated with cave sites dated to 15,000 years B.P. (Bogucki 1996)
Sunflower	*Helianthus annuus*	Moist rocky microhabitats in deserts (Sauer 1993)
Sesame	*Sesamum indicum*	Southern Sahara (Zohary and Hopf 1993)
Cottonseed	*Gossypium barbadense*	Sea cliffs in Ecuador (Sauer 1993)
Olive	*Olea europaea*	Unfavourable sites on mountain slopes in the Mediterranean (Naveh and Vernet 1991)
Hemp	Cannabis sativa	Mountain slopes, human waste sites (Zohary and Hopf 1993, Löve 1992)
Cabbage	*Brassica oleracea*	Sea cliffs, rocky beaches, open habitats (Sauer 1993)
Tomato	*Lycopersicon esculentum*	Coastal deserts (Sauer 1993)
Cucumber	*Cucumis sativus*	Upland rocky slopes in the Himalayas (Zohary and Hopf 1993)
Zucchini, squash, pumpkin	*Cucurbita pepo*	Rocky valleys and escarpments in Mexico (Sanjur et al. 2002)
Peppers	*Capsicum annuum*	Open and disturbed habitats, mountains (Sauer 1993)

Onion	*Allium cepa*	Rocky talus slopes in central and western Asia (Hanelt 1985, Zohary and Hopf 1993, Sauer 1993)
Carrot	*Daucus carota*	Mountain talus slopes throughout Europe and Asia (Ellenberg 1988). Camp follower of human settlements (Löve 1992)
Watermelon	*Citrullus vulgaris*	African desert (Zohary and Hopf 1993)
Cantaloupe	*Cucumis melo*	Dry habitats in subtropical Asia (Zohary and Hopf 1993)
Grape	*Vitis vinifera*	Rock walls, gorges, canopies opening to light (Zohary and Hopf 1993); Mediterranean mountain slopes, unproductive land (Naveh and Vernet 1991)
Date	*Phoenix dactylifera*	Gorges, wet rocky escarpments, desert, open sunny sites, swamp margins, oases (Zohary and Hopf 1993)
Rhubarb	*Rheum palmatum*	Steep rocky talus slopes and hillsides in Nepal (Rock 1930, 1931)
Sugar cane	*Saccharum officinarum*	Tropical stream banks (Sauer 1993)
Beet	*Beta vulgaris*	Rocky shores, banks, temperate forest (Simmons 1976, Zohary and Hopf 1993)
Apple	*Malus pumila*	Fruit forests on talus slopes in western Asia (Juniper 2000)
Pear	*Pyrus communis*	Sunny, rocky slopes, banks of streams (Zohary and Hopf 1993, Sauer 1993)
Plum	*Prunus domestica*	Open woods and cleared hillsides (Zohary and Hopf 1993)
Orange	*Citrus sinensis*	Steep slopes in the Himalayas, China, and northeastern India (Barigozzi 1986, Bonavia 1890)
Lemon	*Citrus lemon*	Steep slopes in the Himalayas, China, and northeastern India (Barigozzi 1986, Bonavia 1890)
Lime	*Citrus aurantifolia*	Steep slopes in the Himalayas, China, and northeastern India (Barigozzi 1986, Bonavia 1890)
Grapefruit	*Citrus paradisi*	Steep slopes in the Himalayas, China, and northeastern India (Barigozzi 1986, Bonavia 1890)
Avocado	*Persea americana*	Cloud forest in Honduras (Barigozzi 1986)
Mango	*Mangifera indica*	Seasonal tropical forest (Simmons 1976)
Banana	*Musa sapientum*	Natural open habitats (Sauer 1993)
Strawberry	*Fragaria* species	Permanently open woods or slopes (Zohary and Hopf 1993, Sauer 1993, Marks 1983)
Almond	*Prunus amygdalis*	Rocky slopes and escarpments (Zohary and Hopf 1993, Sauer 1993); mountain slopes in the Mediterranean region (Naveh and Vernet 1991)
Pistachio	*Pistacia vera*	Dry steppe, slopes, talus (Zohary and Hopf 1993)
Chestnut	*Castanea sativa*	Woodlands on steep slopes (Zohary and Hopf 1993)
Coffee	*Coffea* species	Understory of tall trees, forest clearings (Sauer 1993)
Flax	*Linum usitatissimum*	Seepage areas on rocky outcrops, moist grass (Zohary and Hopf 1993)
Fig	*Ficus carica*	Rock crevices, gorges, stream sides, terrace walls, cave entrances (Zohary and Hopf 1993). Associated with Mediterranean rocky slopes and unproductive sites (Naveh and Vernet 1991)
Pomegranate	*Punica granatum*	Rocky stony slopes (de Candolle 1964), Mediterranean mountain slopes, unproductive land (Naveh and Vernet 1991)
Cherry	*Prunus avium*	Open deciduous woodland (Sauer 1993)
Lettuce	*Lactuca sativa*	Sand dunes, rocky slopes, walls (de Candolle 1964, Sauer 1993)
Coriander	*Coriandrum sativum*	Open forest, steppe (Zohary and Hopf 1993)
Cranberry	*Vaccinium macrocarpon*	Acid bogs (Sauer 1993)
Blueberry	*Vaccinium angustifolium*	Sand, bogs, open upland rocky outcrops (Sauer 1993)
Peanut	*Arachis* species	Seasonally dry savanna, scrub forest (Sauer 1993)
Alfalfa	*Medicago sativa*	Open sites, grasslands, slopes (Sauer 1993)
Agave	*Agave* species	Rocky slopes, cliffs, deserts (Sauer 1993)
Caper	*Capparis spinosa*	Rock outcrops, cliffs (Davis 1951)

Vertebrates are rare among species that form mutualistic relationships with humans, but the few that exist have resulted in massive benefits to both participants. They include species recruited as draft animals, such as elephants, horses, asses, llamas, goats, aurochs, and cattle, as well as those, such as pigs and dogs, that were kept for their ability to eat our refuse organic matter. Besides being kept as pets, dogs and cats were also used to protect people or property, and virtually all of the above plus pigeons were exploited by humans for their feces. In all these cases, the mutualism is based on the animal receiving food and protection from the human, while the human benefits from labour or, in the case of pets, emotional support.

Our mutualistic relationships with invertebrates such as the cockroach are usually viewed with disgust in human societies, but in fairness to the scurrying insects, they are simply taking advantage of our refuse, which would otherwise accumulate and harbour pathogenic fungi or bacteria during its decomposition. Despite the fact that we regard them as pests, dust mites, house spiders, and other species of small invertebrates may serve the same beneficial functions for us. Numerous species of bacteria live mutualistically in our guts; the bacterium *Escherichia coli*, for example, is dependent on us for food and protection and in turn assists us in the uptake of water and vitamin K. Bacteria in our oral cavities benefit from the food that we offer them daily and deter the development of pathogenic bacteria.

Plants that form natural mutualistic relationships with humans include a wide variety of species used in horticulture. Among them are members of the rose family, chrysanthemums, lilies, tulips, petunias, as well as decorative woody plants such as *Cotoneaster*, *Cornus*, *Sambucus*, and *Forsythia*. Trees are often used mutualistically by humans to provide shade, windbreak, or protection from predators. All these examples are of plants exploited for their beauty or for their effect on modifying our living environment.

Our relationships with the many organisms that we cultivate to harvest or eat would be classified by ecologists somewhere on the continuum between mutualism and predation, depending on the relative degree of harm vs. benefit that the organisms derive. We have simply grouped these organisms together in part (c) of Table 1 as the species exploited in agriculture. Two facts are truly remarkable about the animals and plants that humans exploit in this way. The first is that of the estimated 1.5 to 10 million species on the planet, only several hundred have been recruited into agriculture.[34] The second is that each of these species is spectacularly more abundant now than it would have been if the interactions with humans had not taken place.[35] So even though we act as predators on these species, we have certainly been responsible for their massive population increases, too.

Five animal species dominate the agricultural systems of the world (Table 1c): cattle, pigs, chickens, horses, and sheep. Some of these are exploited for their milk, eggs, or fleece, and all of them are eaten by humans in whole or in part. In addition, there are a number of regionally important species that include vertebrates (such as turkeys, goats, guinea pigs, and fish) as well as invertebrates (for example, shellfish and bees).

Table 1 (c) makes it clear that many more plant species than animal species are used in agriculture. Note that the table lists only the species that have *historically* been exploited in agriculture, and that even more species, mostly tropical ones, have been recruited into the human food chain over the past 50 years. Regardless of this, only three species, all of them grasses, dominate the agricultural systems of the world: wheat (Figure 8), rice, and corn together account for half of the global food production, each contributing about 17% of the total. Rye, oats, millet, sorghum, and barley account for another 30%, mostly to feed the animals we eat. The remaining 20% of world food production is divided among a large number of crops that include members of the bean family, the squash family, potatoes, tomatoes, onions, lettuce, cabbage, beets, carrots, celery, parsley, cane sugar, apples, bananas, grapes, and many more (Table 1c). The dominant species used in agriculture recur from continent to continent,[36] but the minor species reflect a regional component that provides for the cultural base of the food systems of the world.

We have now reviewed the plant and animal species that were closely associated with our ancestors and those that are associated with us today as commensals, mutualists, or species we exploit in agriculture. Clearly, we have been surrounded by the same, or a very similar, flora and fauna for several hundred thousand years. The organisms that surrounded our cave-dwelling ancestors would be a familiar sight to all of us, and our ancestors in turn might readily recognize most of the items in our current diet. It is also clear that the species that have been closely associated with humans throughout their history are a small but consistent subset of the entire flora and fauna.

Figure 8.
Field of wheat (*Triticum aestivum*) in Guelph, Ontario. Wheat, like many of the edible grasses, originated on sloping rocky ground in upland areas of the Middle East and western Asia. Photo: P. Kelly.

Where did these organisms come from?

So where did bedbugs live before there were beds, and cockroaches before there were apartment buildings? Where did the asparagus, potatoes, and carrots you ate for dinner or the apples, pears, and rhubarb you had for dessert grow before there were agricultural fields? In many cases, the species we live with or their close relatives may still be found wild in sites that reflect their original distribution. This isn't to say that a feast of wild carrot (*Daucus carota*) would be very pleasant. The wild forms of many of our exploited species have not been selected for traits that make them palatable.

The irritating bedbug is probably the best and most unlikely place to start. This species evolved in rocky habitats in Africa and Asia as ectoparasites of pigeons and bats that lived in caves. They still exist in these habitats, where they hide in cracks and crevices of the rock and wait for their hosts to rest. Blood meals are then taken, and the insects return to their crevices. To the bedbug, the crevices must be stable and close to the food supply, which is easily accomplished in a rocky site with resting bats or birds close by. Humans simply provided a different, larger blood source, and the built structures that followed the cave-dwelling phase simply increased the form and number of crevices for the bedbugs to hide in. The German cockroach evolved, and still exists, on talus slopes in Africa where it scavenges organic debris falling from the cliffs above. As a generalized detritus feeder, it was able to take advantage of the large amount of debris that was associated with human dwellings at the bases of cliffs. As humans moved away from caves to built structures, the cockroach moved with them and eventually spread to Europe along with humans. Asparagus is originally a plant of rock outcrops and talus slopes in Asia and the Mediterranean. The famous 19th-century botanist Alphonse de Candolle reported that William Hooker found wild potato growing on cliffs in coastal Peru. Wild carrot is known to be a camp follower of early human settlements and is still widespread on mountain talus slopes throughout Europe and Asia. Cliff and cave sites in the Blue Mountains of eastern Australia were occupied by aboriginal humans for the period from 22,000 years ago to recent times; in these areas both bogong moths (*Grotis infusa*) and daisy yams (*Microseris lanceolata*) were endemic, and both species became staples of the aboriginal diet.[37] We can trace many other associations between humans and plants or animals back to the time when early humans shared rocky sites with these organisms.

How general is this pattern? To find out, we have consulted the literature on each of the species shown in Table 1 to find evidence for their likely habitat of origin. If no such evidence could be found, then the habitat is listed where the species or its closest relatives are currently found in the wild. The results of our search make it clear that a wide variety of habitat types have given rise to such organisms, but that talus slopes and other rocky areas are listed far more often than any other habitat type. Besides asparagus, potato, and carrot, the plants from rocky habitats that have been recruited into agriculture or horticulture include various members of the rose family, onions, tomatoes, peppers, cucumbers, as well as many garden flowers and shrubs such as carnations (genus *Dianthus*), spurge laurels (*Daphne*), primroses (*Primula*), *Anemone,* poppies (*Papaver*), *Rhododendron, Viburnum,* onions (*Allium*), stonecrops (*Umbilicus and Sedum*), groundsels (*Senecio*), rock cresses (*Arabidopsis*), catchflies (*Lychnis*), tulips (*Tulipa*), *Geranium, Forsythia, Spirea, Cotoneaster, Iris,* Peony, and *Petunia*.[38] Current Chinese agriculture owes much of its existence to organisms recruited from rocky escarpment lands,[39] and Cox (1945) has pointed out that most of the economically important plants that originated in China have come from regions of limestone towers and generally steep topography in that country. Even for the agriculturally and horticulturally important genus

Citrus (the oranges), whose initial date of recruitment into agriculture cannot be determined, there is some early evidence from China that the natural habitat was steep slopes along mountain paths.[40] R. Farrar's 1917 two-volume account of travels through Tibet[41] also adds weight to the idea that cliff faces, rock outcrops, and scree slopes are the prime habitat for a colossal array of plants that are currently important in horticulture. His account is typical of many published in the early part of the 20th century in providing, in exquisite prose, a step-by-step travelogue of his movements through the rocky valleys and what it felt like to see these familiar plants in their natural habitats.

The number of weed species that are derived from persistently open habitats is too great to allow a comprehensive treatment here. The question of the origin of these commensal (or sometimes highly competitive) field plants in Europe, Asia, and North America has been considered by a wide variety of authors,[42] and most workers acknowledge that such plants were once endemic to stream margins, marshlands, beaches, dunes, cliffs, scree slopes, and other bare and exposed rocky sites.

Most people will admit to having no idea where our plant and animal pests, weeds, and domesticated species originated. Ecologists might immediately say "disturbed habitats," without thinking further that such habitats were created by humans much too recently for the species to actually have evolved there. Table 1 provides only the faintest glimmer of how different the reality of where these species arose is from the conventional view. It appears that permanently open rocky habitats including cliffs, rock outcrops, and talus slopes are by far the most common habitats of origin for a very large number of species, both plant and animal, associated with humans. The full significance of this finding will be discussed later in this chapter. First, let us turn in more detail to the stories of a few selected organisms that are familiar to us.

Pigeons

Most people don't think of pigeons as wild animals. They think of them as pests, and in many cities a substantial amount of money is spent trying to get rid of them. In the United States, the figure is 1.1 billion dollars annually.[43] But the case of the pigeon, or more correctly, the rock dove, is both typical and exceptional at the same time. Simms and others[44] suggest that rock doves (*Columba livia*) were endemic to rock outcrops and cliffs throughout much of northern Africa, the Mediterranean, and east to India. By 7500 years ago, rock doves were well-established commensals with

THE URBAN CLIFF REVOLUTION

humans in Mesopotamia and the Mediterranean. Remains of wild rock doves dated in excess of 300,000 years have been found in Israel, and images of pigeons were included in pottery from the Halafian period (8000 to 7000 years ago).

Figure 9.
Pigeon or Rock Dove (*Columba livia*) nesting in the cavity
of a steel girder of a shopping complex in Toronto. This species fully exploits
the capacity of humans to supply it with food. The Urban Cliff Hypothesis
argues that humans have been very effective in providing the same
opportunities for many other species. Photo: P. Kelly.

The biology of the pigeon still reflects its rock outcrop origins (Figure 9). Pigeons only nest on protected ledges of rock or under overhangs that are well protected from the elements. Nests are built up over long periods of time, resulting in considerable accumulations of guano around them. The flight behaviour of pigeons also permits them to fully exploit rock outcrops. Takeoffs can be vertical and explosive; the wings are powered by extremely strong muscles, and the arc of the wing sweep approaches 180° (Figure 10). While gliding, pigeons have a very steep dihedral angle (the angle between the wing and the true horizontal), which increases the inherent stability of the bird's flight, especially in turbulent airstreams. This could be the direct result of natural selection, since rock outcrops have extremely turbulent wind patterns.[45]

Pigeons are largely granivorous but consume a wide variety of food items and can be viewed as opportunistic granivorous scavengers. Their vertebrate predators include cats and raptors such as kestrels, peregrine falcons, barn owls, and eagle owls. Nest location precludes predation by most cats, but chicks are taken from the nest by nocturnal feeding owls such as barn owls. Predation on airborne adults by raptors is also common.

Figure 10.
Pigeons are remarkably powerful and stable flyers. They are seen here roosting, hovering, and exploding vertically into the air. The birds are evidently capable of fully exploiting the food surpluses adjacent to people.
Photo: P. Kelly.

Dogs and cats

The rapid evolution of domestic dogs (*Canis familiaris*) from wolves (*Canis lupus*) is well documented, except that the *location* where domestication occurred has never been appreciated. We believe that the location is most important, and recent work by Coppinger and Coppinger[46] supports this view. Prior to 13,000 years ago, wolves were exploiting caves and rock outcrops for dens and refuge sites and most certainly would have come into contact with humans that were using the same habitat. A consistent pattern of rock shelter exploitation by both wolves and humans at the same time is well established from the fossil record. Based on the evidence of habitat use, we believe that the wolf-human connection was well established for tens of thousands of years before the rapid transition to dog became evident. Canids with dog-like form appeared very suddenly around 13,000 years ago at archeological sites around the world,[47] including Kalamakia cave in Greece[48] and caves in Britain.[49] For the human–wolf interaction to lead to the evolution of dogs (and much more will be presented on this topic later in the book when we write about the way architecture opened up opportunities for domestication), all that was required was natural selection for the willingness on the part of wolves to exploit human-generated refuse and not flee when approached by humans. This selection would have been strong because human campsites in rock outcrops were used continuously for many generations, thereby consistently increasing the ability of the dog progenitors to exploit human waste and increase their own evolutionary fitness by producing more offspring.

The case of domestic cats is less easy to argue because the mutualism between cats and humans is less symmetric: humans throughout history have needed cats more than they have needed us. Even today, feral cats are much more likely to forage freely in wilderness areas far from human settlements than feral dogs, which are rarely found living completely independently of humans. Part of this may be explained by the solitary nature of cats, but part is undoubtedly due to their status as obligate predators for whom the abundance of food items that humans inadvertently make available around their dwellings would hold less attraction than it would for scavengers such as dogs.

The common housecat *Felis domesticus* (Figure 11) is well known to have evolved from the African and European wildcat, *Felis silvestris*.[50] Many varieties of this cat have been described, but all can interbreed, suggesting that varietal status is weakly maintained. The animal can still be found in the wild state in Scotland (where it was introduced by the

Romans), in several countries in mainland Europe, and in northern Africa where it inhabits rock outcrops that are located within or adjacent to forest or desert habitats. *Felis silvestris* skeletons have been found in caves dating back 300,000 to 500,000 years,[51] suggesting that the exploitation of rock outcrops by cats for hunting and refuge may be extremely old.

Figure 11.
A domestic cat (*Felis domesticus*) lurks behind a stone wall in Norwich, England. Cats are behaviourally different but genetically indistinguishable from the European and African wildcat (*Felis silvestris*). The two species behave similarly in feral form. Photo: P. Kelly.

By most accounts, domestication of the cat took place approximately 5000 to 3000 years ago along the rocky banks of the Nile Valley in Egypt.[52] Fossils found at Kalamakia Cave, Greece,[53] show a rapid transition from wolves, ibex, and wildcats in mid-Pleistocene to dogs, domestic goats, and cats in more recent deposits. The transition is more difficult to prove for cats than for dogs because the morphological differentiation following domestication in *Felis* is more difficult to show. Regardless, the sizes of the

THE URBAN CLIFF REVOLUTION

cat populations and their contact with humans undoubtedly increased by many orders of magnitude about 5000 years ago, at the time that humans were constructing outbuildings for grain storage from annual or biannual harvests.

Despite the difference between the progenitors of cats and dogs, they shared some important characteristics that favoured domestication. First, both wolves and wildcats defecate outside the dwelling site rather than in their den. Second, as with wolves (see next chapter), wildcats were rarely acting as predators on humans. Of all the top carnivores in east Africa and the Mediterranean region, *Felis silvestris* is among the smallest and least threatening to humans. Hence, there would have been no selection against humans sharing rocky habitats with wild or domestic cats.

The dependence of ancestral wildcat populations on rock outcrop, cliff, and cave environments probably reflects the benefits that these habitats provide in terms of moderating daily pulses of heat and cold and providing permanent protection from predators. Today, wild populations of *Felis silvestris* are frequently (continental Europe, North Africa including Oman and Libya) or always (Scotland, Germany, France) restricted to high elevation rock outcrop and escarpment settings where the pressures from human disturbance are diminished.[54] This retreat to ancestral habitats in response to ecological stress fits in entirely with the idea of cliffs as refuges described by us earlier.[55]

Rats and mice

Recent work by Austad (2002) suggests that the oldest observations of rats (*Rattus rattus* and *R. norvegicus*) and mice (*Mus musculus*) in close association with humans come from settlements along the Tigris and Euphrates Rivers. These animals are believed to have been migrants from rocky regions at the headwaters of these two rivers. The settlements in question, as we will show later, had an architectural style in which many buildings were intentionally constructed to resemble the cliff dwellings that the humans had occupied previously. Such a linkage is no proof, but it does suggest to us that mice and rats may have originally exploited the same rocky habitats as our human ancestors.[56] Once contact was made, the human waste products and food caches would have represented an added food resource for the rodents. This would have provided an incentive for these rodents to follow the humans as they left the caves and constructed settlements along the rivers 10,000 to 20,000 years ago.

The dependence of rats and mice on rock outcrops and escarpments is clearly not absolute. Rodents such as these represent the most common vertebrate forms worldwide, and wild populations of rats and mice can forage and den in a wide variety of habitat types besides rock outcrops, cliffs, and caves.[57] However, fossil finds in caves confirm that rodents did utilize caves even in prehistoric times, at the same time that these habitats were used by humans. De Lumley (1975) found *Mus* in mid-Pleistocene caves (dating to approximately 800,000 years ago) in southern France, and skeletons of house mice and black rats were reported from Natufian cave dwellings in Israel and Palestine[58] dating to between 15,000 and 12,000 years ago. There are also reports[59] of black rat, house mouse, and gerbils from caves in eastern Asia dating to 100,000 years ago, where rodent remains were associated with the remains of wolves, horses, and pigs (thereby suggesting the presence of humans). Skeletal fragments of *Rattus* and *Mus* were even found with human remains at Olduvai,[60] although caves were not thought to be occupied there. It is irrelevant whether the rodent remains in caves were the result of these species living there on their own accord, as opposed to being transported to the caves in the pellets of barn owls. Both scenarios would clearly indicate a close geographical and habitat-related association between humans and rats or mice.

Livestock

Seventy-five percent of the meat consumed by people comes from just six animal species: cattle, pigs, chickens, sheep, goats, and horses. Of these, only one has a habitat of origin that cannot be described as rocky. The domestic chicken is derived from the red jungle fowl (*Gallus gallus*) that is native to the rainforests of Southeast Asia. All the others originate from rocky habitats in Asia ranging from the steep slopes of the Taurus-Zagros mountains of the Middle East east to China. The various modern varieties of cattle have all evolved from the aurochs (*Bos taurus*). For pigs (*Sus scrofa*), sheep (*Ovis orientalis*), goats (*Capra hircus*), and horses (*Equus caballus*), the modern forms are very similar to the ancestral ones, notwithstanding the effects of selection on body form and colouration. Except for pigs, which are omnivorous, all of these organisms are grazers or browsers of grassy vegetation. Wild forms of these species actively forage on the rocky, open, and highly productive talus slopes at the bases of rock outcrops and mountains. Since no organism is capable of existing without access to drinking water, all foraging takes place in areas adequately supplied with springs, streams, or rivers. None of these animals forms large migratory herds and none is aggressive to humans unless threatened: characteristics in a species that favour domestication.

The various grazers and browsers that co-occurred within a geographical region were segregated vertically by the altitude zone within which they foraged. Aurochs (and later, cattle) grazed the open grassy-filled plains at the bottoms of the slopes. Sheep and goats were capable of exploiting steeper land, rocky scree slopes, and open cliffs at higher altitudes where they were joined by chamois (*Rupicapra rupicapra*) and ibex (*Capra ibex*). North American bighorn sheep (*Ovis canadensis*) and Rocky Mountain goats (*Oreamnus americanus*) still occupy similar zones, although differences in the behaviour and reproductive biology of these two North American species prevented their domestication.[61] The ancestor of the modern sheep, the mouflon (*Ovis orientalis*), lived in rocky talus slope environments and retreated to lower elevations when threatened, mainly because the higher elevations were already exploited by ibex, chamois, and the ancestor of the domestic goat (the mountain goat, *Capra hircus*).

Barrie Juniper (2000) has recently proposed that both pigs and horses evolved in close association with the fruit forests of the Ili Valley in Kazakhstan. These forests occur on well-drained talus slopes of the Tien Shan mountain range, and contain 20 tree species including apple, pear, plum, and apricot. Juniper argues that, in fact, pigs and horses co-evolved with these trees, and that the species were strong agents of natural selection in each other's evolution. A similar co-evolution is thought to have occurred between grazers and grasses in herbivore-plant interactions around the world.

Crop plants

Most people would agree that of all the plant species that we use or consume (Table 1), the one that seems *least* likely to have originated in rocky habitats is rice, simply because of the way it is grown today. However, the current view of rice evolution is that at least the Asian species *Oryza sativa* actually originated on talus slopes in the foothills of the Himalayas, not in permanently flooded valley bottoms as originally assumed.[62] Gorman[63] has produced evidence of 4700-year-old rice grains in caves and suggests that for many centuries before rice was cultivated by humans at lower elevations, wild rice was harvested from natural stands in upland regions and transported to the caves for processing. Chang[64] argues that rice grains so collected were accidentally disseminated into nutrient-rich spots around campsites, where the resulting plants grew luxuriantly. The purposeful cultivation of rice followed, presumably after people observed the greater productivity of the plants when grown in sites that were permanently flooded and supplied with a rich organic nutrient broth. Early commentators[65] did not seem to appreciate that rice was a moist-talus slope grass, but later they drew close attention to beautiful scenes of stunning rock outcrops, stone houses at the foot of the cliffs, mountain streams and rice fields, even noting in one spectacular photograph that such scenes represented models of beauty, prosperity, and peace.

Plants that started their association with humans by accidental dissemination around the campsites, as described above for rice, are often referred to as *camp-following* species. Other camp-followers probably include the ancestor of the large-headed annual sunflower *(Helianthus annuus)*,[66] as well as hemp *(Cannabis sativa)*. Vavilov thought that hemp grew spontaneously in places rich in human organic waste, and was initially tolerated and later actively disseminated by humans because of the benefits the plants provided.[67] These included not only seeds and fibres used for the manufacture of rope, but also their ability to detoxify accumulations of human waste around the campsites.

Mechanisms similar to those applicable for rice have been proposed for the domestication of wheat, oats, rye, barley, and corn. In all these cases, the transportation of the plants from the upland talus slopes where they were collected to the lower-elevation campsites where they were processed was likely a force of natural selection that accidentally produced varieties more suitable for agriculture. Plants whose seeds did not shatter but remained attached to the culm when collected would have been strongly selected for, since their seeds were more likely to be returned to the camp and have a chance to establish near the human habitation site.[68] Hence, selection ultimately produced plants whose grains would not detach on their own accord, making them totally dependent on humans for dispersal.

Wheat *(Triticum aestivum)* in the east and corn *(Zea mays)* in the west are both considered plants of open slopes with limestone soils at low to middle elevations. Close relatives of these species still exist on talus slopes in areas enriched in limestone; their actual ancestors are now known thanks to the use of molecular systematic techniques.[69] Among the many ancestors of domesticated wheat are wild einkorn *(T. monococcum)* and emmer wheats *(T. dicoccoides)*. Both species can still be found growing on the rocky limestone slopes of the Taurus-Zagros mountains in the Fertile Crescent. In a 1925 article in *National Geographic*, Harry Harlan wrote:

Finally, about noon, we came upon a flat bench at the foot of the last cliff. There were numerous fields of emmer, wheat, and barley planted among the boulders. . . . After a brief halt we started up the last unit of the escarpment. It appeared to be only a short distance above us, but the caravan arrived at the crest two and a half hours later, completely exhausted. I photographed the same grainfields from the edge, pointing my camera straight down.

Vavilov[70] was one of the first in the 20th century to visit the sites where the ancestors of grain crops occurred. He did not succeed in his attempts to discover the original habitat of corn in South America, but in Asia, Africa, and Europe he found emmer wheat growing in mountain ravines and in fissures of limestone rocks at modest elevations. Other forage grasses such as rye (*Secale cereale*), oats (*Avena sativa*), and barley (*Hordeum vulgare*) are also well established as plants that grow well on steep limestone soils.[71] Some of their relatives, for example *Hordeum spontaneum*, still occur on rocky escarpments and valley slopes in the Middle East.[72]

The single ancestor of corn is a perennial grass called *teosinte*, now recognized as *Zea mays* subsp. *parviglumis*. It is thought to have evolved in the steep valley of the Balsas River in central Mexico, a place where human-occupied cave sites (such as the Coxcatlan caves)[73] have also been found. Rhoades (1993) mentions that the remains of four-eared corn were discovered in these caves, and such archeological findings are consistent with many creation myths involving corn. While most writers fail to make specific mention of the habitat where wild teosinte occurred, others, such as Barigozzi (1986), state that corn's possible progenitor can be found in open, well-drained, exposed rocky habitats in Central America.

Exactly the same type of site also gave rise to chickpeas (*Cicer arietinum*) and lentils (*Lens culinaris*) in the Old World, and snap-beans (*Phaseolus vulgaris*) and squashes (*Cucurbita pepo*)[74] in the New World. Human-occupied cave sites near rivers in south and central America have yielded the oldest (10,000 years) evidence of New World ancestors of domesticated squash, pumpkin, and zucchini.[75] It is even more surprising that potatoes, cucumbers, figs, pistachios, cabbages, many mustards, olives, almonds, dates, flax, lettuce, spinach, and a wide variety of other garden plants originated in similar habitats as those described above. Hooker was apparently the first European botanist to actually see wild potatoes (*Solanum tuberosum*) growing on cliffs near the sea in Peru.[76] Later, cotton (*Gossypium barbadense*) was also discovered in the same habitat.[77]

Tulips and geraniums

If it is surprising to some that rice – a plant well known to prosper in wet or flooded nutrient rich habitats – actually originated in upland rocky terrain, one might reasonably ask what other familiar plants that require abundant moisture and lots of organic matter might also have origins in rock outcrop habitats. As it turns out, there are lots of them. Some of the most interesting and surprising examples are species that are familiar to the home gardener but which were once endemic to naturally occurring cliffs, rock walls, and talus slopes.[78] Examples would be tulips (*Tulipa*) and geraniums (*Geranium* and *Pelargonium*).

Both of these groups of plants are now among the most common of ornamental garden plants. Some of the grandest formal gardens of the world exploit massive plantings of tulips and geraniums to achieve the effect of colour explosions that would be unimaginable in natural settings. Since tulip mega-culture originated in Holland, many view the rich wet soils of the lowlands as the natural habitat for the species in this genus. However, a closer look at the garden environment gives a better indication of the habitat from which these species emerged 400 years ago: talus slopes.

A planting bed represents a well-watered, well-drained, often physically confined site with light soil and a high level of organic matter incorporated in the upper layers. Competition from other plants is absent, there is little or no canopy, and there are few predators. Where in nature do such conditions occur? Open habitats that might qualify include river banks, marshlands, beaches, dunes, cliffs, and talus slopes, but most of these lack at least one key feature. River banks are flooded frequently; marshes have lush vegetation and intense competition and predation; and beaches and dunes are very low in organic matter. This leaves cliffs and talus slopes, yet, until they consider it more carefully, most people believe there couldn't be a greater difference than between a cliff or talus slope and a garden.

Solid rock is impossible to colonize for all but a very small minority of plants and animals.[79] However, talus slopes and even cliffs do not consist entirely of solid rock, but rather of rock that is fractured and broken up by natural forces, resulting in a large variety of spaces that can be colonized. Since gravity is constantly removing small particles that might accumulate, little actual soil development takes place. Instead, the spaces collect organic matter that falls from the plateau above, from the rock face, and from plants growing in the spaces themselves. This organic matter accumulates in the upper layers of the spaces where it gets replenished by new litter fall each year as the lower layers decompose. Cliffs and talus slopes are also much wetter than the surrounding landscape, at least in temperate and subtropical latitudes.[80] This is because hydraulic pressures that redistribute percolating rainwater within the bedrock force water toward the cliff face, creating slow seeps that bleed water slowly during the entire growing season and supply the spaces between rocks and the talus with water.

Figure 12.
Many species of Tulip (*Tulipa*) are endemic to rock cliffs and talus slopes of western Asia, even though their incorporation into horticultural beds with nutrient-rich soils might indicate otherwise. Photo: A. Nejedly.

Talus slopes and fractured rock faces, therefore, support real "rock gardens" in which plants have access to full radiation, water, nutrients, and reduced competition and predation – a description of the traditional tulip garden (Figure 12).

This revelation is best put into context with some particularly vivid passages from an account by Pavord[81] of her encounters with wild tulips and geraniums in eastern Turkey:

[The Renault] carried us deep into the bare hills and rocky screes where pockets of bright red tulips grew among small geraniums and eremurus just coming into flower … The loveliest colonies of tulips we found were in a valley above Tortum, north of Erzerum where groups of *T. armena* grew in little pockets between limestone crags. We always found something intriguing there, sometimes a draba, sometimes an iris, once a wolf. That day, I was spread-eagled with my eyes closed, on a flat piece of rock in the sun. The *T. julia/T. armena* conundrum was rolling around my head like a riddle. I opened my eyes – who knows why – to find a wolf silhouetted against the sun. It sat upright, facing me on a neighbouring rock, its tail neatly curled around its front legs. Only inches from my eyes were the tulips, brilliant red blazes in the foreground. Behind them was the wolf, stark against the sky. When I sat up, it bolted away, disappearing into a low cave under a neighbouring rock crag. The conjunction of the two was as enigmatic in its way as the saints had been in Crete. As I lay on in the sun above Tortum, I thought still of these tulips, slashes of brilliant blood welling from the bare, brown, shale-strewn slopes of the mountain. Wolves were nothing to them. Saints were nothing to them. Millennia had passed by this slope, while the tulip, wild as the wolf, slowly, joyously had evolved and regenerated itself. Even now, in their dark underground grottoes beneath the rocks, the tulips were plotting new feats, re-inventing themselves in ways that we could never dream of.

Likewise, if somewhat more concisely, Wilson (1946) records:

There on the southwest coast in the bay of Angra Peguena [Africa] in a chasm of white marble rock, …[Antoni Pantaleo Hove] found a thick-stemmed, brand-new geranium with roots several yards long. They were naked and hard as wire and the heat was so

THE URBAN CLIFF REVOLUTION

intense on the rocks as to blister the soles of the feet. Yet this lovely white-flowered, round-leaved *Pelargonium crassicaule* in mid-April was blooming there as freely as if it were growing in a well-tended English garden. The plants appeared to have received their nourishment solely from the moisture lodged there during the rainy season, assisted by a little sand drifted by the wind into the cavities. … It is from just such hardy tolerant ancestors that today's geraniums draw their strength.

Some of the most remarkable text concerning the origin of tulips comes from the detailed species accounts given by Pavord. For *Tulipa aleppensis* she writes:

Hall grew it in a cool greenhouse and remarked on its habit of "wandering by stolons". These crept some way from the place where the original bulb was planted and lodged bulbs into crevices of the greenhouse wall.

For the massive and brilliant red *T. fosteriana*:

In the mountains south of Samarkand . . . it grows in deep soil among limestone rocks at about 1,700 m [elevation].

And for *T. lanata*:

[It] occurs in Kashmir, where it is an introduction and grows on the roofs of temples and mosques.

Thirty of Pavord's habitat descriptions for 55 tulip species mention rocky scree slopes as the habitats of origin. We are not saying that all tulips are rock outcrop specialists, but rock outcrops as habitats for these plants clearly show up far more frequently than expected.

There are far too many examples of familiar plant and animal species that had their origins in rocky habitats for us to tell stories about all of them. Instead, we will go on to ask just what habitats did we *expect* the organisms that we have recruited into our lives to originate from?

Surprising origins: nurseries of stone

Should we be surprised to find that a large proportion of the plants and animals associated with us throughout history are native to rock outcrops, cliffs, and talus slopes? To answer this question, we will summarize the habitat variability on earth and then show how vastly overrepresented rocky habitats are (see Table 1), given their commonness among the earth's terrestrial habitats. In subsequent chapters we will try to explain why this might be so by considering the features of each habitat that could have been exploited by humans in order to satisfy their basic needs during the long period of our evolution.

Global habitat variability
Tropical rain forest and tropical seasonal forest represent roughly 20% of the total vegetated land surface on earth,[82] while temperate forests (including evergreen, deciduous, and boreal forest) account for another 19%. Open woodlands, shrub lands, and savannas represent 19%, temperate grasslands 7%, tundra and alpine areas slightly more than 6%, and desert and near-desert 14%. All wet habitats such as lakes, rivers, and marshes combined make up 3%. Cultivated land currently accounts for the balance of slightly more than 11%.[83]

These biomes, as they are called, are not equal in their ability to sustain plants and animals. Tundra and desert support an average of 0.7 kg of plant biomass per square meter, and the new growth accumulated each year averages only 0.09 kg m^{-2} (Table 2). Grassland, savanna, and open woodland are almost an order of magnitude more productive. Their biomass averages almost 4 kg m^{-2}, with an annual increment of about 0.7 kg m^{-2} per year. Forests are vastly more productive even than grasslands and savannas: all forest types combined have an average biomass of 33 kg m^{-2}, producing an annual increment of about 1.4 kg m^{-2} per year.[84]

There is no question that major fluctuations in the relative land area taken up by different biomes have taken place since the beginning of the Pleistocene. For example, the amount of wet forest was reduced and the amount of steppe, grassland, tundra, and desert was increased during periods of climatic cooling.[85] Note, however, that while cliffs, talus, and rocky habitats can occur in any of these biomes, rocky habitats, as we will show in the next section, account for no more than a tiny fraction of the total vegetated landscape area.

TABLE 2

The ability of different habitat types on earth to satisfy the energy needs of humans. Each of the different biomes (habitat types) is listed with the total percentage of vegetated land surface that it occupies (after Lieth and Whittaker 1975). For each biome, the average standing crop (plant biomass per square meter) and productivity (new growth accumulated per square meter and per year) are given as cited by Lieth and Whittaker (1975), and the percent allocation of productivity to edible matter (seeds or edible storage organs) is given as cited by Harper (1977). From these values we have calculated the production of edible matter per square meter per year for each biome, and the area that would be required to satisfy the energy needs of one human (400 kg of carbohydrates per year).

Biome	Total % of vegetated land surface	Average standing crop (kg m^{-2})	Average productivity (kg m^{-2} yr^{-1})	Allocation to edible matter (%)	Production of edible matter per year (kg m^{-2} yr^{-1})	Area required to satisfy energy needs of one human (hectares)
Tundra and desert	20	0.7	0.09	<1	0.0009	44.6
Grassland, savanna and open woodland	26	4	0.7	30–50	0.21	0.19
Tropical and temperate forests	39	33	1.4	1–5	0.014–0.07	2.8
Cliffs	probably<1	2–4	0.04–0.06	unknown; assuming 1–30	0.00004– 0.0069	95–290

Surface area of cliffs

The surface area of cliffs has not been evaluated on any continent or for any biome, mainly because such estimates are usually obtained by remote sensing, and remotely sensed images give cliff faces a vertically projected area of zero. In this section we want to convey a sense of how little cliff habitat is available for organisms to exploit compared with the amount of other habitat types on the planet. We accept that we may be artificially reducing our estimate by excluding from consideration unvegetated land areas such as extreme desert, mountain tops, sand, and ice, many of which have a high concentration of cliffs. However, we would like to limit our argument to rock outcrops and cliffs that are located at lower elevations within otherwise productive biomes.

Let us first consider a hypothetical example that demonstrates the relationship between the apparent commonness of cliffs and their actual surface area. Imagine the most extreme case of a landscape riddled with cliffs (Figure 13). This hypothetical landscape area measures 10 km x 10 km, or 10,000 hectares. Cliffs occur in 10 rows, each 10 km long, adding up to a total length of cliff face of 100 km. There are very few places on earth with such a high density of cliffs. If we assume the cliffs to be 25 m high, a normal modest size for sedimentary rock, then the total surface area of vertical cliff face would be 25 m x 100 km, or 250 hectares. This represents 2.5% of the horizontal landscape area. If the cliffs are of a more substantial height (50 m), then the area of cliff face is 500 hectares, or 5% of the landscape; if the open faces are extremely high (100 m), then the cliff face area is 1000 hectares, or 10% of the landscape.

This hypothetical scenario might perhaps apply to Grand Canyon National Park in the United States or to the fissured valleys of Jordan (Figure 14), but it is hardly representative of most other areas on earth. If we replace the 100 km length of open cliff per 10 km x 10 km area by a more realistic value of 10 km, and if we allow for some of the cliffs to be discontinuous or partly mantled, the estimate of the surface area of vertical cliffs drops one or two orders of magnitude to well below 1% of the landscape area.

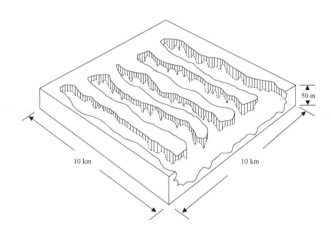

Figure 13.
Diagram illustrating the relatively small surface area of cliff faces even in a landscape riddled with cliffs. In this example, a 10 x 10 km grid has 100 km of open cliff face averaging a face 25 m in height. There are few places in the world with such a large concentration of cliffs, yet the surface area of the exposed free-face is still only 250 hectares within the 10,000 hectare total area. Adjacent talus slopes may represent another 250 hectares.
Illustration: D.Larson.

THE URBAN CLIFF REVOLUTION

Consider some real-life estimates for this from actual landscapes. The Niagara Escarpment in southern Ontario includes an estimated 350 hectares of open cliff face. The cliffs are located within a protected corridor of relatively undisturbed landscape, the Niagara Escarpment Planning and Development Area, of 196,100 hectares. Even within this zone that has cliffs as the central and most protected feature, the area of vertical cliff face is only 0.18% of the horizontal land area. Since cliffs do not occur commonly throughout the rest of Ontario, it follows that their area is vanishingly small when expressed on the basis of the land area of the province as a whole.

In Haute-Provence, southern France, in an area enriched in limestone rock outcrops and canyons around the Verdon Gorge, vertical cliffs 10-600 m in height extend for a total of 62 km within a 20 km x 20 km area (40,000 hectares). If we assume an average cliff height of 100 m, we can calculate the total area of vertical face as 620 hectares, or a landscape fraction of 1.55%. In a second area of France south of Grenoble, a 10 km x 10 km area east of Chabestan contains 10 km of open cliffs 50 m in height, representing 0.5% of the landscape.

Another area enriched in limestone cliffs can be found in the Peak District, Derbyshire, U.K. Approximately 20 km of cliffs averaging 25 m in

Figure 14.
A fissured limestone valley with cliffs in Jordan. Photo: J. Lundholm.

height can be found there within a 20 km x 20 km block of land centred on Matlock-Bath, amounting to 0.125% of the horizontal area.

Many more examples make the same point, among them a 9300-hectare National Park called Sächsische Schweiz in Germany that appears to be densely packed with tall sandstone towers. Even there, the 70 kilometres of cliff face averaging 50 m in height only amount to 3.7% of the park area.

As far as talus slopes are concerned, their area will be at least as great as the surface area of the cliffs they are associated with[86] but can also be greater in situations where the cliff face has fully broken down. For the sake of argument, we will assume double the area for talus slopes as for open cliff faces. All these values are only rough estimates, but cliffs and talus slopes represent only a vanishingly small proportion of the total surface area of otherwise productive biomes. This is true even for landscapes where cliffs are common, such as the canyon lands in the western United States or the Great Rift Valley of east Africa.

Not only is the area taken up by rocky habitats small, but their ability to produce biomass is also very small. For the Niagara Escarpment in Ontario, we have calculated that the biomass of lower plants (algae and lichens) per square meter of rock face amounts to 0.0015-0.07 kg and has an annual productivity of 0.004 to 0.02 kg per square meter.[87] We estimate that higher plant biomass on the cliffs makes up no more than 2–4 kg m^{-2}, and considering that the growth rates of trees on the Niagara Escarpment [88] and other cliffs around the world[89] are known to be vanishingly small, the resulting total productivity of the cliff habitat is probably extremely low.

A test of an idea

From the proportion of available habitat on earth that is rocky and from the global patterns of terrestrial biomass and productivity, we might have predicted that forests and grasslands – the most abundant and productive habitats on earth – should have produced the majority of the plants and animals that are associated with humans. Cliffs, talus slopes, and other rocky landforms represent only a tiny proportion of the total land area and biomass within any biome; we might have expected, therefore, that very few species would have been derived from these places.

In Table 3, we have used values for the total area of different habitat types on earth to calculate how many of the species listed in Table 1 would be expected to come from each habitat type if they all contributed equally. We then contrast this with the number of species actually derived from each habitat. The difference between what is expected and what is observed is huge. The number of species in Table 1 that have cliffs, talus slopes, or rocky habitats as their habitat of origin is several orders of magnitude larger than the expectation based on the availability of such habitats. In fact, more than half of the species listed in Table 1 originally lived in rocky habitats, even though we could have reasonably expected only one such species on the list. Perhaps an even bigger surprise is that forests, representing 38.8% of the terrestrial surface, have supplied only eight species on the list. This anomaly is even more conspicuous when one considers that tropical forests are among the most species-rich habitats on the planet. True, a great effort has been made in the past 30 years to recruit many tropical forest plants into the human food chain, but these efforts have in many cases followed extensive experimentation in plant breeding to make the plants palatable. The minuscule number of species that had been recruited into agriculture up to then is truly surprising.

TABLE 3

The origin of the plants and animals that interact with humans as derived by quantitatively summarizing the data in Table 1. All habitat types that are referred to in Table 1 were assigned to the habitat categories of Lieth and Whittaker (1975). For each habitat category, the total number of times it occurs in Table 1 was then tallied (number of entries) and was also expressed as a percentage of the total number of entries in the table. Species for which multiple habitat types were listed in Table 1 were permitted multiple entries. The percentage of the earth's surface that is taken up by each habitat category is given after Lieth and Whittaker (1975). Based on this percentage we have calculated how many of the entries in Table 1 would be expected to fall into each of the habitat categories if organisms recruited into agriculture had been derived with equal probability from all habitat types (% habitat type x total number of entries in Table 1).

Habitat category (after Lieth and Whittaker 1975)	Number of entries in Table 1	% of entries in Table 1	% of area taken up on earth	Expected number of entries in Table 1
Tropical rain forest and tropical seasonal forest	5	4.8	19.6	20
Temperate forest (evergreen, deciduous or boreal)	3	2.9	19.2	20
Woodlands, shrublands, savanna	12	11.6	18.8	19
Temperate grasslands	10	9.7	7.2	7
Tundra and alpine regions	1	0.97	6.4	7
Desert and near-desert	10	9.7	14.4	15
Lakes, rivers, or marshes	8	7.8	3.2	3
Talus and slopes	34	33.0	<<1	<<1
Caves and cliffs	20	19.4	<<1	<<1

TULIPS, PIGEONS, AND PEOPLE

So we are left with two conclusions at this juncture in our investigations. First, that many of the plants and animals that are associated with us – whether they are our pets, pests, or food items – appear to have originated in rocky cliff or talus slope environments. Second, that this is completely unexpected considering that rocky habitats are both exceptionally rare and exceptionally unproductive in comparison with neighbouring forest and savanna habitats.

---

17   De Candolle (1964, reprinted from original 1886 edition). De Candolle is considered by many to be the father of research into the origins of plants and animals used in agriculture. He was the first to suggest that the valley of the Tigris and Euphrates was an important region. Later work, however, showed that the plants and animals found there were brought in from other regions prior to the establishment of cities and towns along the rivers.
      Löve (1992), translation of Vavilov (1926). This translation is interesting by itself, but to understand the full contribution of N.I. Vavilov to the science of domestication, read Popovsky's 1984 volume describing the persecution of Vavilov by Lysenko and Stalin.
      Baker and Stebbins (1965). This book emerged from an early symposium on the topic of plant invasions and the characteristics of weeds.
      Simmons (1976).
18   Samuels and Prasad (1994). This comprehensive volume examines the interactions between humans and their "built" environment.
19   For example, read David Quammen's article, "Planet of Weeds," in *Harper's* (October 1998). Also read his earlier article on pigeons (Quammen 1996).
20   Spirn (1984). A classic work in the field of urban landscape architecture.
21   Ward (2002). A fascinating mix of well-accepted views on evolution by natural selection, and ideas about where this evolutionary process might be going.
22   Kahlke (1975), Binford (1984).
23   Jaeger (1975), Potts (1984).
24   Smith (1992). See also the work by Coppinger and Coppinger (2001). They suggest that it is very difficult to separate wolves from early dogs based on skeletons alone.
25   Lavaud-Girard (1993).
26   Zeuner (1963).
27   Pokines (2000).
28   See, for example, Golovanova et al. (1999) and Prasad (1996).
29   Carrión and Scott (1999).
30   Wenke (1990), Beadle (1977).
31   Barigozzi (1986).
32   Buckler et al. (1998).
33   For starlings see Cabe (1993) and Schneider (1990); house sparrows, Lowther and Cink (1992);

barn owls, Brown and Brown (1995).
34 Wiersema and Leon (1999).
35 Budiansky (1992) argues that "domesticated" plants and animals have actually exploited us as much as we have exploited them. Budianski points out that some species such as the horse would probably have followed companion species to extinction in the late Pleistocene if the association with humans had not taken place.
36 FAO World Food Production, vol. 52, 1998.
37 Bird et al. (1988), Helms (1890), Argue (1995).
38 Farrar (1917); Hanelt (1985); Wu and Raven (1996); Fairchild (1919). Ando and Hashimoto (1998) and Sink (1984) show that *Petunia* in particular is considered to be a species that colonizes moist and nutrient-rich river banks. But these descriptions are for South American rivers that cut through rock strata forming steep-walled canyons with talus slopes that merge directly with the river. Ando and Hashimoto point out that the actual microhabitats where their new species of *Petunia* grow are exposed spaces between rocks on these sloping canyon walls.
39 Ping-Ti (1977).
40 Han (1178). This volume existed as a manuscript and was translated into English and published in 1923 by M.J. Hagerty.
41 R. Farrar (1917).
42 Liebman et al. (2001), Baker (1974), Marks (1983), and Delcourt (1987). The Marks paper is important because it is one of the first to indicate that rock outcrops and cliffs represent permanent waste places where certain species that can be considered as weeds evolved.
43 Pimental et al. (2000).
44 Simms (1979) and Johnston and Janiga (1995). Simms explicitly notes that the natural ancestral habitat and the modern urban habitat of pigeons are one and the same.
45 Larson et al. (2000a). Readers should spend some time actually watching pigeons fly around buildings. They are amazing birds.
46 Coppinger and Coppinger (2001). This volume is a "must read" for anyone interested in the process of domestication.
47 Morey (1994).
48 De Lumley and Darlas (1994).
49 Smith (1992).
50 Randi and Ragni (1991). The genetic similarity of these taxa suggests (as does the work with dogs/wolves) that many species barriers are entirely behavioural and may be controlled by immeasurably small genetic differences.
51 Gamble (1999a).
52 Kurtén (1965a, b), Clutton-Brock (1981).
53 De Lumley and Darlas (1994).
54 Easterbee et al. (1991), Hubbard et al. (1992), Kitchener and Easterbee (1992).
55 Larson et al. (2000a).
56 Bunney (1994).
57 Ranck (1968), Feldhammer et al. (1984).
58 Thorpe (1996).
59 Aigner (1978a,b) and Higham (1977). Regarding the former study, all of the fossil remains of rodents associated with human remains at four South African cave sites and at several sites in Spain (Pokines 2000) were obtained from regurgitated owl pellets formed by barn owls (*Tyto alba*) over the middle to late Pleistocene (Avery 1995). Likewise at Westbury Cave, U.K., most of the small mammal fossils were obtained from owl pellets produced by cliff-dwelling owls such as the barn owl and the European eagle owl (Andrews 1990).
60 Binford (1984).
61 Carr (1977).
62 Chang (1976).
63 Gorman (1977). The transportation of the rice to the caves does not prove that the rice was growing nearby, but it is highly likely.
64 Chang (1976). This reference, and the one below, both argue that the widespread cultivation of rice followed a long period of time in which it naturally grew in moist, nutrient-rich habitats that were well drained. The cliff-face Buddhist shrine known as the Eagle's nest (Bhutan) is immediately above rice fields that are nestled in at the bases of the talus slopes.
65 Groff and Lau (1937).
66 Heiser (1965).
67 Löve (1992).

68   Ryan and Pitman (1998). This volume, entitled *Noah's Flood*, does not focus on the issue of plant domestication, but in describing how campsites followed the retreating margins of the glacial lake (that is now the Black Sea), the authors present the scenario of wheat ancestors being harvested from distant talus slopes and transported back to the camps.

69   Doebley (1990).

70   Löve (1992), translation of Vavilov (1926).

71   Sauer (1993).

72   El Hadidi et al. (1986).

73   Salvador (1997).

74   Sanjur et al. (2002).

75   Smith (1997).

76   De Candolle (1964), reprint of 1886 edition.

77   Sauer (1993).

78   Stace (1999).

79   Cryptoendolithic algae, cyanobacteria, and lichens can live below the surface of solid rock (Matthes-Sears et al. 1997).

80   Larson et al. (2000a).

81   Pavord (1999). This is an amazing volume, filled with information and beautiful illustrations.

82   We are considering here only the 125,000,000 km² of the earth's land surface that is currently vegetated. The 24,000,000 km² that is glacier, open rock, or sand supporting little or no vegetation will be excluded from consideration. The total productivity of these habitats is so low and the total number of species they support is so small as to have made little or no contribution to the array of plants and animals associated with people.

83   Lieth and Whittaker (1975).

84   Lieth and Whittaker (1975).

85   Gamble (1999a).

86   Larson et al. (2000a).

87   Matthes-Sears et al. (1997).

88   Kelly et al.(1992).

89   Larson et al. (2000b).

CHAPTER THREE

Gimme Shelter:
What We Needed
and Where We Found It

Many people in Western societies make a habit of claiming that they *need* things like a second car or a CD player. These claims of need fuel our consumer society and are responsible for the allocation of vast amounts of the world's resources to a few Western economies. Sanderson and co-workers (2002) have estimated that as of the year 2002, human beings have directly or indirectly influenced 83% of the terrestrial surface of the earth. Much of this exploitation provides for wants, not needs. Try this little experiment. Imagine that trees had never been harvested for lumber and bricks had never been manufactured. Imagine also that right now, at this very second, all built structures disappeared and that clothing did not exist. Where would you sleep tonight? What would you *need* to survive until the morning? How comfortable would you feel, and what is the relationship between that comfort and your needs?

These questions may seem silly, but for a vast number of animal species including most grazing mammals and birds this scenario is very common. Sufficient food and water would certainly help by providing the energy needed to endure the wind, cold, and rain, but you would have to expend considerable energy to obtain food and water in the first place. The risk of death is greatest when hypothermia is most likely. Most warm-blooded vertebrates deal with this risk by moving to warmer locations (short-term solution) or growing body coverings (long-term solution). Another obvious solution is to kill a large furry animal and use its skin for shelter. (This might sound unappealing, because we're used to furriers making the practice appear elegant and sophisticated.)

We have presented this scenario not only to show how vulnerable we are to the vagaries of the physical environment out of doors, but also to emphasize that *comfort* is something we can live without. The list of real human needs includes energy and nutrients, water, shelter from the physical environment, from predators, and from resource fluctuations, tools, knowledge, and spiritual resources. We will examine these in more detail in this chapter, but first we will briefly review the history of human evolution and human dwelling sites. Knowledge of the history of the habitats that people sought will make the list of human needs more understandable. The reason for proceeding this way is simple: the idea is well established in people's minds that humans are descended from tree-dwelling ancestors who then migrated to savanna habitats as the climate changed. The problem with this conventional understanding is that it leaves certain questions unanswered. Such as: where did we live and how did we both enjoy the benefits and survive the threats of savanna living at the same time? We think that rock played a central role in this new habitat.

A short history of human evolution

In the conventional view of human evolution, our primate ancestors once lived in trees and then, over a two-million-year period, moved onto the savannas of east Africa. There they developed bipedalism, became hairless, ate everything in sight, and colonized land like crazy. New varieties then developed a capacity to make tools and control fire. These forms persisted for several hundred thousand years. Finally, there appeared a smarter, more handsome version of the same creature that we would not mind calling an ancestor. This form was artistic, sensitive, mobile, innovative, and capable of producing computers that never crashed . . .

This story is one of the world's favourite natural history topics and has been so thoroughly treated by a large number of experts that we do not need to present a comprehensive review of it here. Most aspects are still being actively researched,[90] and intense controversy remains over some of the evolutionary connections between our ancestors. For example, the recent discovery in Portugal of the 25,000-year-old skeleton of a young boy with features that are intermediate between *Homo neandertalensis*[91] and *Homo sapiens* sparked numerous insoluble arguments over the species name that ought to be used for this single and unusual specimen.[92]

For the purpose of this book, we will present a general overview that sets the stage for linking human evolution with the habitats of origin of our companion species. We will briefly outline the sequence of human ancestors that appeared over the past million years or so, the timing of their appearance and disappearance from the fossil record, and the features of the environment within which they lived.

Human evolution began in the Pliocene period with the appearance of the forest-dwelling species *Ardipithecus ramidus* roughly 4.4 million years ago. It was followed by *Australopithecus anamensis* (4.2 to 3.9 million years ago), *Australopithecus afarensis* (3.6 to 3.2 million years ago), *Australopithecus africanus* (3 to 2 million years ago), and finally the first rock tool user *Homo habilis* (roughly 2 to 1.5 million years ago; Figure 15a). All of these hominids were forest or forest margin dwellers,[93] and their use of the savanna biome increased only at the end of the Pliocene. These species are relevant to the Urban Cliff Hypothesis because they frame the transition from hominids dependent on trees within open woodlands, river margins, and forests to those dependent on rock outcrops and cliffs within savannas.

The Pleistocene Epoch, which followed the Pliocene from 2 million to 0.7 million years ago (Figure 15b), is generally regarded as a cooling period during which forests were gradually replaced by grasslands in central Africa and Asia. Three species of *Homo* were contemporaneous, including *H. habilis*, *H. ergaster*, and *H. erectus*. The last is thought to have ultimately given rise to *H. antecessor*, the putative ancestral species to modern humans.[94]

Unlike the early Pleistocene, the middle and late Pleistocene periods were marked by intense episodes of glaciation. During the middle Pleistocene, starting about 700,000 years ago, there was pronounced variability in climate with at least eight major and several minor cooling trends (Figure 15c). In this period, *Homo neandertalensis* in Europe was contemporary with *Homo erectus* in Africa and central Asia. *H. erectus* is thought to be the first human species to be able to control the use of fire; the first and last appearance dates of this species are roughly 1.7 million and 250,000 years ago, although some recent work[95] suggests a much more recent date for its final extinction in Indonesia.

The late Pleistocene (Figure 15c) brought a cooling trend that started about 127,000 years ago and was interrupted by two brief warming spells, each lasting a few thousand years: one 105,000 years ago and another 82,000 years ago.[96] In this period, both *H. neandertalensis* and *H. sapiens* lived in Europe and around the Mediterranean. The earliest *H. neandertalensis* fossils date back about 300,000 years, and the last Neanderthals are believed to have perished in Western Europe about 27,000 years ago.[97] Anatomically (though not culturally), modern specimens of *H. sapiens sapiens* (which we will call simply *H. sapiens* here)

Figure 15.

A four-panel illustration showing the change in hominid species, the changes in climate, and the changes in cultural development over the past 4 million years. This figure summarizes information gathered from Tattersall (1995, 1998), Gamble (1986, 1999a), and Vrba et al. (1995). Illustration: D. Larson.

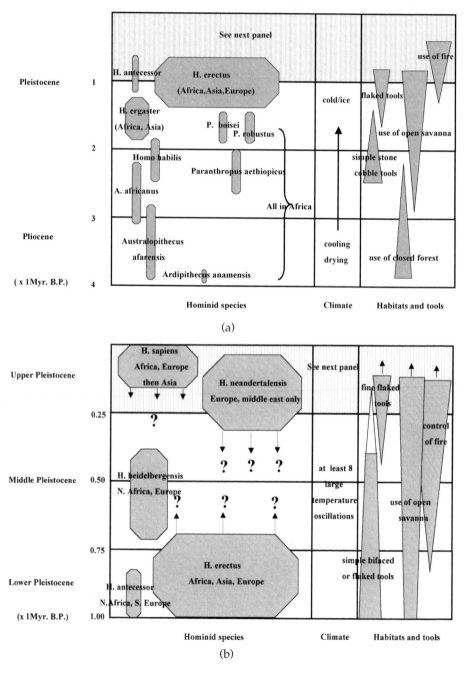

Panel (a) covers the period from 4 to 1 million years before present (B.P.); (b) from 1.0 to 0.25 million years B.P.; (c) from 250,000 to 50,000 years B.P.; and (d) from 50,000 years B.P.

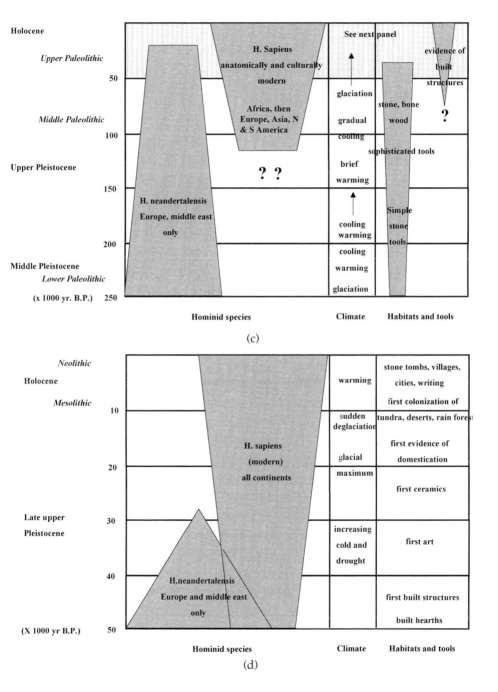

(c)

(d)

may date from as early as 100,000 years before the present[98] (Figure 15c, d).

Between the last glacial maximum (roughly 18,000 years ago) and the beginning of the Holocene (the last 10,000 years), dry grasslands rather than forests persisted in places where humans lived. Cold and drought were occasionally replaced by moderately warm and wet conditions, but not for extended periods.[99]

Global warming and deglaciation have occurred throughout the Holocene, and modern humans were living on all continents by this time. The colonization of the Americas, however, is thought to have taken place only 11,000 to 12,000 years ago.[100] The first evidence of agriculture dates to between 12,000 and 6000 years ago, depending on the continent in question.[101] During this period, the label *Mesolithic* is applied to people who were hunting, gathering, and farming, whereas the label *Neolithic* is applied to people with better developed agriculture. Farming spread farther during the modern historical time period, which begins with the smelting of metals such as copper and bronze.

Where did humans live throughout prehistory?

We will show in the following section that cliffs, talus slopes, caves, and rock shelters were exploited by early humans at rates that greatly exceeded the commonness of these landscape elements in nature. We have already shown, in the preceding chapter, that such habitats represent an exceptionally small amount of land relative to the area available as forest, grassland, wetland, or tundra. To what extent were these different habitat types used by humans?

The search for evidence of past habitat use suffers from two major problems: the relatively low overall frequency of human remains and tools, and the fact that not all habitats preserve the evidence of use equally. When human specimens are located, it is difficult to know to what degree the evidence may have been changed by the physical processes operating at that site. Some sites may enrich while others may deplete the evidence, resulting in what are called *taphonomic biases*. For example, evidence of human settlements on river banks may be carried downstream to the lake basins during floods, giving a false indication of the location of the occupation. Most archeologists are aware of the degree to which site conditions remove the signs of habitat use and resource exploitation and try to interpret the evidence that remains.[102] But in order to make our argument, we must consider the possibility that any patterns found are artifacts of differential preservation. We must acknowledge at the outset another source of bias inherent in the use of published evidence: that the authors of these works themselves may have been selective in the habitats they have investigated. Keeping these potential pitfalls in mind, we will now review the published archeological literature that describes the sites where major groups of fossils and tools have been found. We will then summarize the conclusions that paleoanthropologists have drawn from the available evidence about the patterns of habitat use by early humans.

The ancestors of modern humans were originally forest dwellers, but a gradual shift to more open habitats occurred during the late Pliocene and early Pleistocene. *Australopithecus anamensis* preferentially exploited forested sites, while the later *A. afarensis* and *A. africanus* tended to inhabit more open woodlands and the edges of savannas.[103] The consistent exploitation of savanna began with *Homo habilis* between 2 and 1.5 million years ago (Figure 15a).

Both recent[104] and early[105] accounts of the evolution of *Homo* from its *Ardipithecus* and *Australopithecus* ancestors emphasize that climatic change played a large part in the habitat shift. The cooling and drying of Africa and Asia during the late Pliocene and early Pleistocene was accompanied by a reduction in the total area of woodland and forest and an increase in the extent of open woodlands, savanna, and grassland.[106] We will show later in this chapter that in comparison with forests, savanna and grassland offer a much greater amount of edible biomass to granivores, herbivores, and carnivores. Consequently, there must have been strong selection for the forest-dwelling great apes, whose growth rates and population sizes were probably limited by low food availability in the forest, to exploit more and more of the resources of the savanna biome. The forces of selection to exploit savanna, however, would have applied not only to early humans but also to their competitors and predators. Accordingly, strong selection pressures would also have operated on the choice of refuge sites, especially in species that lacked physical strength such as *Australopithecus* and *Homo*. We have established elsewhere[107] that cliffs and rock outcrops represent such refuge sites and are exploited as "enemy free space" by a wide variety of vertebrates including rodents, marsupials, certain birds, and many primates. Of course, cliffs are not restricted to the savanna biome, but as Africa and Asia became cooler and dryer, such habitat features became increasingly valuable as refuges for species whose principal defence mechanisms against predation and the physical environment were camouflage or hiding. Some modern reports[108] state that the use of caves as rock shelters is not restricted to hominids, suggesting that some primates may be predisposed to be able to exploit these microhabitat elements efficiently. Precise numbers are impossible to glean from the literature, but we believe that there is evidence for a steady increase in the use of rock resources and rock shelters along with the increased use of savanna by early humans during the late Pliocene and early Pleistocene (Figure 15a). Pfeiffer (1972) writes:

The arboreal life was not yet completely abandoned, however. At first the [early hominids such as *Ramapithecus*] may have continued sleeping in trees among the woodlands where they returned every night. Later, when they no longer returned to the woodlands, they slept in groves out in the savanna. When trees were not available they most likely chose places where predators could not follow, perhaps spending their nights on the ledges of cliffs, facing the cliff wall as some baboons do in the twentieth century. . . .

Early hominids, like their fellow primates, lived mainly exposed to the elements. Sometimes on stormy nights, however, they may have been driven to seek cover under overhanging cliffs, where they huddled together wet and cold in the dark in a world still dominated by other animals. Sometimes bad weather may even have driven them into caves. Early one morning in South-west Africa during a very cold, harsh season, [K.] Hall saw a troop of baboons emerging from a cave high in a cliff overlooking a river bed, and on another occasion he took motion pictures of a troop leaving a cave by the edge of the sea. But habitual cave dwelling came later. Caves were occupied by more efficient killers which had to be driven out and kept out, and that had to await the widespread use of fire. On the other hand, brief visits may have been common, a possibility suggested by the finding of hominid remains, and bones presumably shattered by hominids, at Swartrans and other Transvaal caves. . . . Hominids may have entered these caves in the daytime to obtain water and get out of the noonday sun rather than for extended periods of shelter.

Pfeiffer makes a valuable point, but we believe he has underestimated the degree to which cliffs were used by early hominids. The significance of cliffs, caves, and rock shelters is rarely emphasized in the paleoanthropological literature. However, accounts of field observations at the sites where human remains are preserved make mention of such habitats with striking frequency. For example, Dart's first report (1925) of *Australopithecus africanus* was based on a specimen obtained from a cave deposit at the base of a limestone cliff near the village of Taung in South Africa. At the Sterkfontein site, also in South Africa, Robert Broom found the first remains of *Paranthropus robustus* in a karstic limestone tower; other fossil-yielding sites have been reported from caves in the nearby Makapan Valley in an area with open cliffs and waterfalls.[109] One of the first reports of *Homo erectus* at Choukoutien in China was based on 600,000-year-old fossils retrieved from a debris-filled cave at the base of large limestone tower.[110] A general consensus has now developed that hominids living in Southeast Asia during the period from 1.5 to 0.5 million years ago exploited upland caves as dwelling sites.[111] Similar rock shelters in southern alpine France (Le Vallonnett, Roquebrune-Cap-Martin) were inhabited by *H. erectus* between 1.7 million and 700,000 years ago.[112] There are many more examples: again and again throughout the Pleistocene fossil record, sites with familiar names that gave rise to ancestral *Homo* are described as caves at the base of limestone cliffs or escarpments.[113] The trend continues through to the Holocene and anatomically modern humans. Some open-air sites have been found in the British Isles, but the vast majority of the evidence for long-term settlements there comes from caves at the bases of escarpments or cliffs.[114] Examples of these include Kent's Cavern, Creswell Crags, Gough's Cave at Cheddar Gorge, Mount Sandel (Ireland), MacArthur Cave, Ulva Cave (Mull), and Cnoc Coig. Caves were still being used in Britain by anatomically and culturally modern humans 25,000 to 15,000 years ago.[115] The same caves were occupied for thousands of years, but not necessarily on a continuous basis. In fact, good evidence from British as well as German and French sites indicate that the occupation was seasonal.

Similar evidence can be found in many other parts of the world. In North America, recent cave occupation sites are found in the southwestern deserts, through the plains, and into Illinois and Ohio; the Modoc Rock Shelter in Illinois is an example.[116] Anatomically modern humans also occupied caves in the Tehuacán Valley in Mexico, the Andean highlands, and Australia.[117] Kempe (1988) has documented that cave dwellings were once common at sites such as Spirit Cave, Non Nok Tha (Thailand), and still occur today in many parts of central and southern Asia. They are even featured in Greek mythology: Zeus was born in a cave, and Sibyl lived in a cave under the Temple of Apollo. And according to Byzantine belief, Jesus Christ was born not in a stable but in a limestone cave.[118]

Other habitat types such as clay pits or swamps are less frequently mentioned as the sites where human fossils were discovered, but this cannot of course be taken as evidence that such open-air sites did not exist. Indeed, many workers indicate that remains of humans exist in open sites as well as cave sites.[119] However, it is surprising how frequently cliffs and rocky habitat are mentioned as the locations where human remains are found. (Table 4). This is especially so when one considers that the evidence of the actual refuge or shelter is easily obscured by physical weathering and destruction of the sites. Figure 16[120] illustrates how evidence of caves or rock shelters can easily be lost by the time human remains are discovered. In east Africa, volcanic eruptions during the late Pliocene would have added accumulations of ash and other debris to any rock shelter sites, making detection even more difficult. Even without taking this into account, human remains are found in sites associated with cliffs far more frequently than one would expect considering the frequency of these elements in the landscape.

One of the main arguments used to dispel the idea that humans used cliffs more than other habitat types is the differential preservation of bony remains in caves as compared with open sites. There is no doubt that the microclimatic and geomorphological conditions within rock shelters were particularly favourable for the retention of fossilized materials such as bone,[121] and caves could be enriched in bony remains simply because the material was preserved there while most remains from open sites were lost. However, stone tools, which are not subject to decomposition, are also found more often in caves than in open sites; from this we may conclude that cliffs and rock outcrops were actively sought by early humans. If the relative scarcity of open-air remains is due to the dissolution (or other loss) of bony materials, then one would further expect evidence for open campsites to become much less frequent over time. In fact, sites that were occupied only 10,000 years ago show the same trend toward the exploitation of rock shelters by humans as do much older sites (Table 4). Oddly, researchers have readily taken the presence of cave bear, cave lion, and cave hyena skeletons in caves as evidence that these species exploited caves,[122] while showing great reluctance to accept the same reasoning for people. Why people would rather believe that early humans mainly exploited open-air sites, even though there is little fossil evidence for this, may be best explained by the cartoon in Figure 1. Note that Gamble (1994)

TABLE 4

A survey of the literature characterizing the sites where human remains were found. For each reference, the following is given: species of hominid found; approximate date of site; geographical location and name of site (or number of sites if more than one); habitat setting where the remains were found; species of animals and plants whose fossils were associated with those of the hominids; and other artifacts present at the site. Entries are ordered by approximate age of findings within each geographical region.

Hominid species	Approximate date (years B.P.)	Location	Habitat	Associated animals	Associated plants	Artefacts present	Reference
Europe							
Homo erectus	1.4 –0.7 million	Southern France, (Vallonett Cave)	cliff/cave	dog, cat, bear, pig, cattle, deer, rhinoceros, horse, mouse, porcupine, garden dormouse	none listed	tools	de Lumley 1975
Homo erectus or early *H. neander-talensis*	0.7 million	France, (Cave of Escale)	cliff/cave, hillsides, river banks, loess plains	dog, cat, bear, pig, cattle, deer, rhinoceros, horse, mouse, porcupine, garden dormouse	none listed	tools	de Lumley 1975
Homo erectus	~800,000	Europe (general)	cliff/cave	cave bear, lion, wolf, fox, pig, deer, reindeer, chamois, goat, bison	none listed	tools	Butzer and Isaac 1975
Homo erectus	400,000– 200,000	France	cave	bear, lion, hyena, wolf, fox, pig, deer, reindeer, chamois, goat	none listed	tools	Kahlke 1975
Homo neander-talensis	unknown	Germany (46 sites)	outcrops	elephant, cave bear, hyena, wild cat, beaver, wild boar, roe deer, hedgehog	none listed	tools	Lyell 1873
Homo neander-talensis	unknown	Italy (Palermo Bay)	cliff/cave	cattle, deer, pig, bear, dog, wolf, cat	none listed	tools	Lyell 1873

Hominid species	Approximate date (years B.P.)	Location	Habitat	Associated animals	Associated plants	Artefacts present	Reference
Homo neandertalensis and *H. sapiens*	200,000 - 20,000	Balkans	cliff/cave	none listed	none listed	tools	Bunney 1994
Homo neandertalensis	130,000	France (numerous sites)	cliff/cave	hyena, wolf, fox, hare, horse, boar, deer, mountain goat, cattle, chamois	juniper, alder, grasses, rock polypody, spikemoss, composites	tools	Bordes 1972
Homo sapiens	100,000– 60,000	Spain	cliff, cave	none listed	none listed	tools	Kraybill 1977
Homo sapiens	80,000 –40,000	U.K.	cliff/cave and open sites	none listed	none listed	tools	Gamble 1994
Homo sapiens	40,000 –20,000	Germany (4 sites along Danube)	cliff/cave	pig	none listed	tools	Mithen 1990
Homo neandertalensis	38,000 –30,000	France (Saint-Césaire)	cliff/cave	reindeer, mammoth, rhinoceros, horse, cow, pig, megaceros, deer, wolf, fox, cave hyena, bison, chamois, goat	juniper, pine, alder, rock polypody	tools	Lévêque 1993
Homo sapiens	30,000	France	cliff/cave	none listed	none listed	art	Clottes and Courtin 1994
Homo sapiens	30,000 –20,000	France	cliff/cave	none listed	none listed	art	Chauvet et al. 1996
Homo neandertalensis and *H. sapiens*	25,000	Portugal	cliff/cave	none listed	none listed	tools	Kunzig 1999
Homo sapiens	15,000 –10,000	France	cliff/cave	none listed	none listed	art	Kahlke 1975
Homo sapiens	15,000	Balkans	cliff/cave, river margins	none listed	emmer wheat, einkorn wheat, barley, lentil, pea	none listed	Bogucki 1996
Homo sapiens	12,000 –10,000	U.K (14 sites)	cliffs, caves	hyena, brown bear, cattle, horse, rabbit, wolf, dog, peregrine falcon, stock dove, rock dove,	chamomile, meadow rue, dandelion, juniper, willow, crowberry, buckthorn, oak, elm	tools	Smith 1992

Hominid species	Approximate date (years B.P.)	Location	Habitat	Associated animals	Associated plants	Artefacts present	Reference
North America							
Homo sapiens	10,000	Illinois	cliff/cave	none listed	none listed	tools	Caldwell 1977
Homo sapiens	10,000	Missouri (numerous sites)	cliff/cave	none listed	none listed	tools	Johnson 1980
Homo sapiens	6000–2000	South-western U.S.	cliff/cave	sheep, horse, pig, cattle, rabbit, wolf, woodrat	juniper	tools	Carr 1977
Homo sapiens	4000	Belize (198 sites)	cliff/cave	none listed	none listed	tools	McNatt 1996
South America							
Homo sapiens	11,000	Peru	cliff/cave	horse, ground sloth	none listed	tools	MacNeish 1977
Homo sapiens	11,000 –1500	Peru	cliff/cave	guinea pig, puma	potato, common bean, Lima bean, maize, peppers	tools	MacNeish 1977
Homo sapiens	9000 – 6,500	Peru	not specified	guinea pig, dog, llama	none listed	tools	Wing 1977
Homo sapiens	7,000	Peru	cliff/cave	none listed	teosinte	tools	Beadle 1977
Asia							
Homo erectus	0.7 million	China (Chou-koutien)	cliff/cave	none listed	none listed	tools	de Lumley 1975
Homo erectus	700,000 –300,000	south China	cliff/cave	pongo, rat, bear, dog, wolf, horse, pig, deer, mouse	none listed	tools	Aigner 1978a
Homo (species not identified)	200,000	India	cliff/cave	none listed	none listed	tools	Prasad 1996
Homo sapiens	100,000 –30,000	south China	cliff/cave	stegodon, hyena, horse, pig, cattle, porcupine, rhinoceros	none listed	tools	Aigner 1978a
Homo neander-talensis	53,000 –30,000	south China	cliff/cave	none listed	pea, almond, betel, gourd, pepper, water chestnut, bean	tools	Clark 1977

Hominid species	Approximate date (years B.P.)	Location	Habitat	Associated animals	Associated plants	Artefacts present	Reference
Homo sapiens	50,000	India (numerous sites)	cliff/cave	none listed	none listed	tools	Sankalia 1978
Homo sapiens	40,000 to present	Philippines (Tabon and 5 other sites)	cliff/cave	none listed	none listed	tools	Fox 1978
Homo sapiens	35,000	Borneo	cliff/cave	none listed	none listed	tools	Harrisson 1978
Homo sapiens	28,000 –4,000	Philippines (Pilanduk Cave, Guri Cave)	cliff/cave	pig, deer	none listed	tools	Fox 1978
Homo sapiens	19,000 –10,000	Southeast Asia	cliff/cave	elephant, rhinoceros, cow, goat, deer, elk, pig, bear, tiger, porcupine, squirrel, mouse, rat, bat	rice (not cultivated)	tools	Gorman 1977
Homo sapiens	14,000 –8,000	Cambodia	cliff/cave	deer, langur, macaque, gibbon, civet, badger, rat, squirrel, pig, serow, and goral (both goat-like endemic ungulates)	none listed	tools	Higham 1977
Africa							
Australopithe-cus boisei	1.5 million	Eastern South Africa	riverside	rat, mouse, pig, antelope, frog	none listed	tools	Leakey 1960, Potts 1984
Homo erectus	500,000	South Africa (numerous sites)	cliff/cave	none listed	none listed	tools	Deacon 1975
Homo ergaster?	400,000	South Africa	cliff/cave	none listed	none listed	tools	Clark 1975
Homo neander-talensis	100,000	Southern South Africa (Klasies River)	cliff/cave	rat, mouse, pig, antelope, frog	none listed	tools	Binford 1984
Homo sapiens	46,000 –42,000	Africa	cliff/cave	none listed	none listed	tools	Kraybill 1977

Hominid species	Approximate date (years B.P.)	Location	Habitat	Associated animals	Associated plants	Artefacts present	Reference
Homo sapiens	33,000	Africa (Olieboom-spoort Cave)	cliff/cave	none listed	none listed	tools	Kraybill 1977
Homo sapiens	15,000	Africa (Transvaal)	cave	none listed	none listed	Tools	Kraybill 1977
Homo sapiens	15,000	Africa (Mwulu)	cave	none listed	none listed	tools	Kraybill 1977
Homo sapiens	5,000 –3,000	Northern South Africa	cliff/cave	cattle	oak, cypress, olive	tools, art	Lajoux 1963
Homo sapiens	2000	South Africa	cliff/cave	none listed	none listed	tools	Solomon 1996
Australasia							
Homo sapiens	60,000	Australia	cliff/cave	none listed	sweet potato, grasses	tools	Gamble 1994
Homo sapiens	30,000 –10,000	Australia /Tasmania	cliff/cave	none listed	none listed	tools	Gamble 1994
Homo sapiens	26,000– present	Papua New Guinea	cliff/cave	none listed	none listed	tools	Gamble 1994

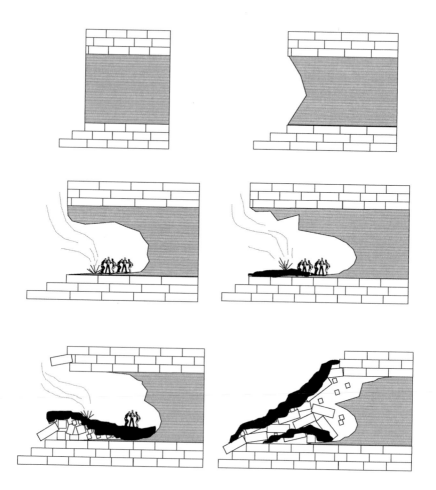

Figure 16.
Illustration of the stages in the development of rock shelters and caves in sedimentary rock. Modified after Bordes (1972). Illustration: D. Larson.

has concluded that humans didn't actually begin to *make* protective structures until about 40,000 years ago. Hence, open-air sites would have been very dangerous for human ancestors to occupy.

Figure 17.
Charles Knight painting of *Homo neandertalensis* in a rock shelter in central Europe. American Museum of Natural History negative number 39441A. Used with permission.

Even though small-scale habitat use is often difficult to reconstruct from archeological evidence, many workers have published accounts of the likely foraging patterns and subsistence activities of ancient humans. Again cliffs, caves, and rock shelters are rarely emphasized but are pictured frequently as an important part of the habitat used by early humans (Figure 17). *Homo ergaster*, the first tool-using *Homo* species to migrate out of Africa into central Asia, is said to have exploited lake margins, stream confluences, and rock outcrops formed by erosion of the river channels as the key components of an optimal foraging strategy.[123] In this matrix of habitat types, the lake margins and stream confluences supplied water, grasses, and herbs to consume and herbivores to hunt. The rock outcrops provided raw materials for construction of tools, as well as shelters from the sun's rays, precipitation, competitors, and predators. Unlike the situation that confronted arboreal primates living in the forest, the four components (water, productive land, a source of tools, and shelter) were all physically adjacent in the landscape. Similarly, Isaac (1975) observes that in the early to middle Pleistocene, foraging bands of humans used river valleys as their base camps because they provided various features that could be exploited: water, forest cover that gave shelter from the midday sun, productive river banks, savanna habitats along plateaus adjacent to river valleys, and rock shelters in places where river systems had cut through rock pediments.

Hundreds of other studies involving investigators in Africa, Asia, and Europe and covering the period from the early Pleistocene through to the Holocene report the exploitation of rock shelters as sites for butchery, tool making, daytime and nightime sleeping and shelter, and protection from cold and rain. Most activity involving the use of fire seems to have occurred near the mouths of caves at the bases of cliffs, most likely because this positioning simultaneously provided light, a reduced amount of smoke, and the ability to keep a vigilant watch on predators. Johanson and Edgar (1996) conclude that the systematic use of such caves continued throughout the Pleistocene into the modern era. The caves were used as home bases for long periods of time and were easy to locate because the cliffs were visible from afar. Johanson and Edgar also point out a little known but revealing fact: that the now less commonly used name for modern humans, *Cro-Magnon*, is actually an old French word meaning "big cliff" and refers to a large limestone massif near Les Eyzies, southern France, a place where more than 200 human campsites have been found. Referring to this region, Pfeiffer (1972) writes:

Shelter, of course, was provided by the cliffs, eroded structures formed by the lime-containing remains of tiny animals deposited and consolidated more than a hundred million years ago in the warm shallow sea that covered most of Europe. The cliffs provided raw material as well as shelter. Embedded in the limestone were large quantities of fine-grained flint in the form of nodules which, like the limestone, consist of the remains of microorganisms (in this case, colonies of single-cell animals with silica containing shells). Water was also available, runoff from the mountains in the Massif Central, where the Dordogne and Vézère rivers rose and joined a few miles below Les Eyzies. . . .

Pfeiffer adds another observation that will become important later in this book:

[Humans] must also have had a feeling for the beauty of the land, and for the hard-won security of standing with a solid wall at his back and looking out over a river valley. From the tops of the highest cliffs one looks into the distance and sees inviting backcountry valleys and other cliffs. . . .

Similarly, Baring-Gould (1911), in beautiful early 20th-century prose, writes:

Evidence substantiating the thesis of [Charles] Lyell had been accumulating, and the researches of Lartet and Christy in the Vézère valley, published in 1865–75, as *Reliquae Aquitanicae*, conclusively proved that man in Perigord had been a naked savage, contemporary with the mammoth, the reindeer and the cave-bear, that he had not learned to domesticate animals, to sow fields, to make pots, and that he was entirely ignorant of the use of metals.

Since then, in the valley of the Vézère, Les Eyzies in the Department of the Dordogne, has become a classic spot. . . . On reaching the valley . . . one is swung down from the plateau into a trough between the steep scarps of chalk-rock that rise 150 to 300 feet above the placid river . . . Out of their clefts spring evergreen oaks, juniper, box and sloe-bushes. Moss and lichen stain the white walls that are streaked by black tricklings from above, and are accordingly not beautiful – their faces are like that of a pale, dirty, and weeping child with a cold in its head, and who does not use a pocket-handkerchief. Jackdaws haunt the upper ledges and smaller caves that gape on all sides chattering like boys escaped from school, and anon a raven starts forth and hoarsely calls for silence.

At the foot of the stooping crags, bowing to each other across the stream, lie masses that have broken from above, and atop and behind these is to be seen a string of cottages built into the rock, taking advantage of the overarching stratum of hard chalk; and cutting into it are russet, tiled roofs, where the cottagers have sought to expand beyond the natural shelter: they are in an intermediate position . . .

Nature would seem to have specially favoured this little nook of France, which must have been the Eden of primeval man on Gallic soil. There he found ready-made habitations, a river abounding in fish, a forest teeming with game; constrained periodically to descend from the waterless plateaux, at such points as favoured a descent, to slake their thirst at the stream, and there was a nude hunter lurking in the scrub or behind a stone, with bow or spear awaiting his prey – his dinner or his jacket.

What beasts did he slay? The wild horse, his huge head, was driven by him over the edge of the precipice, and when it fell with broken limbs or spine was cut up with flint knives and greedily devoured. The reindeer was also hunted, and the cumbersome mammoth enable a whole tribe to gorge itself. . . . The grottoes perforating the cliff, like bubbles in Gruyère cheese, have been occupied consecutively to the present day.

Baring-Gould's statements emphasize the spectacular beauty of the cave and cliff environments exploited by humans.

Despite the overwhelming amount of evidence that cliffs, caves, and rock shelters were essential components of the savanna biome used by early humans including *Australopithecus* and *Homo*, most of the literature on human origins uses only the word *savanna* to characterize the landscape that provided humans with resources. This interpretation prevailed early on in the history of paleoanthropology, when Leakey (1960) described the habitats of early to mid-Pleistocene *Paranthropus boisei* in the Olduvai Gorge as rivers, river banks, and neighbouring savanna, and has not changed materially in the four decades since his work was first published. Some reports[124] mention that most campsites at Olduvai were adjacent to water courses but not so close as to be exposed to floods, but there is no reference to cliffs or rock shelters. Intriguingly, however, references to savanna in the text of a volume are sometimes accompanied by illustrations that emphasize rock outcrops and cliffs. An excellent example of this is a book called *Early Man*.[125] The text as well as the figure caption on page 112 describe the savanna habitat exploited by *Homo erectus*. The caption reads: "Streams, reed-fringed lakes and a warm climate offered *Homo erectus* hospitable campsites in his early East African homeland." But the image shows a habitat entirely different from most people's idea of savanna (Figure 18).

Overall, evidence for the use of rock shelters and caves at the bases of cliffs increases markedly during the middle Pleistocene, the period in which *Homo erectus* declines and *Homo heidelbergensis* increases. The trend continues for both *Homo neandertalensis* and anatomically modern *H. sapiens.* Clive Gamble (1999a) argues that Neanderthals frequently used

Streams, reed-fringed lakes and a warm climate offered Homo erectus hospitable campsites in his early East African homela

Figure 18.
The caption to this illustration reads: "Streams, reed-fringed lakes and a warm climate offered *Homo erectus* hospitable campsites in his early east African homeland." The caption ignores the obvious fact that this "hospitable" campsite is beneath a cliff. Illustration taken from White and Brown (1973). Used with permission.

open sites during the period between 500,000 and 300,000 years ago, but that caves were increasingly exploited from that period through to the beginning of the Holocene. This is supported by recent studies that have successfully extracted sequenceable DNA from Neanderthal specimens retrieved from caves throughout central Europe.[126] In the Fertile Crescent, Paleolithic humans occupied caves and rock shelters of the Taurus-Zagros Mountains and used a variety of animals and plants from around the sites as a food source.[127] This central idea has been captured in some striking images,[128] including murals (Figure 19) painted by Jay Matternes in the American Museum of Natural History.

Figure 19.
Illustration of *Homo erectus* returning fresh kill to rock shelters bordering a productive African savanna. Illustration by Jay Matternes, from pages 64–65, Early Man, courtesy of Time-Life Books and taken from Howell (1965).

What did we need and where did we find it?

So far in this chapter we have summarized what is known about the evolution and habitat use of ancestral humans. The word *habitat* refers to the place where organisms spend their lives, and of course, the time spent living is mainly devoted to acquiring the resources to keep living. In view of the popular idea that humans are a savanna species, not a cave or cliff species, we will now examine the various resources that humans truly *need*. We will then show that a cliff in close proximity to savanna would have been the ideal habitat for ancestral humans to exploit to satisfy these needs.

Throughout this part of the book we have assumed that early humans were not grossly different from human beings today in their biochemistry, physiology, anatomy, or behaviour. Certainly, modern humans are likely to provide a better model of how our ancestors functioned than any other species living today. We have taken great liberty with the literature in order to come up with simple estimates for the parameters that are relevant. We recognize the risks of using such estimates, but our intention is to point out patterns that are important to consider even if inter-individual variation is ignored. We hope to show by the end of the next section that the true needs of our ancestors could not have been satisfied by savanna habitat alone. Instead, the exploitation of rock outcrops and cliffs adjacent to savannas and watercourses is what allowed our species to succeed.

Energy and nutrients

If our Pleistocene ancestors had similar energy requirements to those of modern humans, their resting metabolism would have consumed roughly 6270 KJ[129] (1500 Kilocalories) per person per day. This amount, which is the minimum energy required for basic body functions only, is remarkably constant for modern humans around the world.[130] Greater energy requirements result from physical exertion, as well as from exposure to temperatures away from the zone of thermal neutrality.[131] Exposing a stationary person to subzero temperatures can increase the energetic requirement to 17,000 KJ (~4000 Kcal) per person per day, as can intense physical activity such as running or arguing with one's spouse.

To support an "average" metabolic rate of 11,500 KJ (2750 Kcal) per day, a person would have to consume approximately 1100 grams (dry weight) of carbohydrates or 650 grams of fat each day.[132] On an annual basis, the energy requirement per person would therefore be roughly 400 kg of carbohydrates or 240 kg of fat. Carbohydrates are mainly obtained from plants, but fat can be derived from either plant seeds or animals. In addition to energy, humans require nutrients. These are usually combined with energy in the food that humans consume, although the relative proportions can vary. For example, since meat has a much higher nitrogen content, much less of it needs to be consumed in comparison with plant matter in order to satisfy the human nutrient requirements.

The source of the energy consumed by living humans varies enormously from culture to culture. Strictly vegetarian societies exist as well as those that are strictly meat-eating. Our forest-dwelling Pliocene ancestors were probably omnivorous with a heavy bias toward eating plants and fruits, a diet similar to that of primates living today. The archeological record documents a change in the form of hominid teeth[133] that occurred simultaneously with the increased exploitation of savanna. This suggests that a shift in the hominid diet occurred as meat was more easily obtained by hunting the abundance of granivores in the open grasslands.

Over the past 20 years, a controversy has developed about whether our ancestors depended more on hunting, collecting, or scavenging during our hunting and gathering evolutionary phase. Some evidence does indicate that most of the caloric requirements of early humans were fulfilled by wild plants, but shows that most of the nutrients were supplied by the smaller volume of animal protein obtained by hunting.[134] Some have argued that early hominids could not have relied heavily on carrion[135] because modern humans, like present-day great apes, could not have tolerated the pathological conditions involved with the consumption of carrion due to the morphology and physiology of their intestinal tracts.[136] However, McNeill (1976) points out that such pathogenic microorganisms were normal components of the gut flora of many organisms that were commensal with us, including rats, mice, and pigeons, and that early humans could have gained tolerance of these pathogens by repeated low-level exposure to them over thousands of years.

How do different types of habitats on earth compare in their ability to sustain our energy and nutrient needs? We showed in Chapter 2 that the different biomes are far from equal in the amount of standing crop they support and the new growth they accumulate each year (Table 2). What ultimately determines their ability to sustain humans, however, is not only the amount of biomass that they produce but also how much of it is suitable for consumption by humans or the herbivores that provide food for humans. In the perennial plants that dominate tundra and desert, roughly 90% of the annual productivity is allocated to roots and the remainder to wood, leaves, and fruit. Edible seeds or other carbohydrate storage organs usually account for 1% or less of the annual production, or roughly 0.0009 kg m^{-2} yr^{-1}, and this is enough to support modest numbers of herbivores and carnivores. Grasslands, savannas, and open woodlands are almost an order of magnitude more productive than tundra and deserts. The grasses and herbs that dominate these habitats allocate approximately equal amounts of energy to stems, roots, and seeds, although this can vary widely from species to species with some herbs producing 30% to 50% of their aboveground increment as seeds.[137] Due to the open structure of these habitats, most of the plants do not compete for light, and they are generally not disadvantaged by grazing organisms. In fact, many of the plants that occur in these habitats are tolerant of grazing or dependent on grazers for pollination, seed dispersal, and nutrient redistribution. This feature is important, because it means that such plants are usually non-toxic to herbivores. If 30% of the annual productivity of a grassland or savanna is in the form of seeds or other carbohydrate storage organs such as roots or tubers, this would amount to 0.21 kg m^{-2} yr^{-1} of edible matter that could provide for human energy needs.

The average biomass of all forest types combined is an order of magnitude greater than that of grasslands, and the annual production is twice as high (Table 2). However, more than 90% of the biomass of forests consists of wood, and most of the new growth produced each year is wood or leafy tissue. Trees allocate a much smaller proportion of their annual production to seeds and fruits than grasses and other herbaceous plants, often as little as 1%.[138] While nut producing trees such as walnut, hickory, and avocado might produce individual fruits that are large and fleshy, the total weight of such material per tree is small relative to the amount of new wood, bark, and leaves added each year. These latter plant structures are largely fabricated by plants to increase their competitiveness in a shady environment. Less than 1% of available sunlight usually reaches the ground below forest canopies, compared with more than 30% to 40% in grasslands. Canopy shade has for millions of years placed strong selection pressure on plants not only to be able to compete for position within the canopy, but also to protect themselves against the loss of photosynthetic area. This strong and persistent selection has resulted in the evolution of plant structures that are indigestible, as well as secondary compounds that are toxic to herbivores. Some tropical plants are known to synthesize and reallocate toxic alkaloids the moment leaf tissue is browsed in the canopy. For these reasons, forest habitat, despite its much greater biomass and productivity, can be up to 20 times *less* productive in terms of supporting human life than savanna habitat (Table 2). Anyone who has ever tried to "live out in the woods" without resorting to feeding from lakes will know the difficulty. In fact, most of the French and English explorers of North America who perished in the 16th and 17th centuries did so because they were not able to find suitable food in the forests through which they travelled.

Given the carbohydrate requirements of 400 kg per person per year as calculated above, a human would need to completely harvest and eat the seeds and storage organs of plants from roughly 44.6 hectares of tundra, 2.8 hectares of forest, but only 0.19 hectares of savanna (Table 2). Savanna is thus roughly 20 times more exploitable to humans than is forest. Our conclusion that temperate grassland or savanna was the only habitat easily exploited by early humans is also supported by Gamble (1994) in his book *Timewalkers*.

In the previous chapter we saw that a disproportionately large number of species important to humans are derived from rock outcrops, cliffs, and talus slopes. How do cliffs, these distinct habitats that occur within the confines of other biomes, compare in their ability to support the energy needs of humans? We showed in the same chapter that the total area taken up by these rocky habitats is small and their ability to produce biomass is also very small. The new growth accumulated each year per unit area of cliff face is an order of magnitude lower than for savanna, and more comparable with the productivity of tundra and desert. There are no data available on the allocation of biomass to edible structures for cliff vegetation, but even if we assume it is as high as for savanna (30%), the area of cliff face required to satisfy the energy needs of a single human would still be 95 hectares (Table 2). Clearly, humans could never have sustained themselves by foraging on the biomass of cliffs directly. We propose that the biodiversity of rocky habitats was exploited "off-site": the species originally found here were encouraged to grow elsewhere under human control. We will expand on this notion in Chapter 4, and by the end of the book we will show that one of the biggest risks to the planet is that people are building new cliff habitats (cities) at the expense of the other habitats that fuel our lives.

Water

The amount of water required by humans varies enormously depending on physical exertion and environmental conditions such as temperature and humidity. In modestly vigorous physical activity used in the calculations for energy balance above, a person requires approximately up to 4 liters of water (or equivalents provided by the water content of foods) per day,[139] or 1460 liters per year. Under extreme conditions of low humidity, high wind speed, or high physical exertion, this value can increase to one liter per hour, but usually only for short periods of time.

There is little or no published evidence on the sources of water exploited by early humans. Modern humans derive much of their water from groundwater through the drilling of wells, but there is no known example of such wells being constructed by human ancestors. The direct consumption of precipitation is extremely inefficient (if you were to lie on your back with your mouth open for the duration of a torrential thunderstorm you would still only collect about 150 milliliters of water). However, the runoff from precipitation can be exploited for a limited duration after a storm, or water can be consumed by drinking directly from a lake or river. Another possibility is the consumption of juicy fruit or flesh.

Of these options, only the eating of fruit or flesh involves a source of water that is portable. Primates, unlike camels, have no physiological mechanism for internal water storage, and the first indication of humans' being able to actively cache liquid water did not emerge until well into the Neolithic period, when pottery and skin bags came into use. The lack of transportability of water, its ability to evaporate from storage containers, and the fact that precipitation is very irregular in most habitats other than tropical rainforest all contributed to the need for early humans to position their settlements adjacent to natural water courses such as lakes, rivers, or streams.

We have already determined the amount of land of different habitat types needed to supply the energy needs of prehistoric peoples. Was this minimum area sufficient to simultaneously provide for water needs, or were water needs more limiting than energy needs in certain habitats? To answer this we need only look at the habitat type where water was most likely to be limiting: deserts, a habitat colonized by humans only in the Holocene. Deserts usually receive no more than 250 millimeters of precipitation per year. Each square meter of desert, therefore, directly receives 250 litres of water during an average year.[140] Thus, to provide the 1460 liters of water that an individual person requires per year, only the water arriving at slightly more than 7 square meters needs to be collected for use. The 44.6 hectares required to meet a person's energy needs would most certainly be sufficient to meet their water needs as well, even when taking into account that only a fraction of the precipitation may be harvestable because much of the rainfall occurs in brief downbursts that cannot be exploited for long after the event. The same conclusion can be reached for grassland and savanna. For these habitats, a much smaller area is needed to supply a person's energy needs (2.8 hectares), but much more precipitation falls per year (500–1000 mm of rain), so that the water needs of one person are fulfilled by an area just under three square meters.[141] It is fair to conclude that human existence has been fundamentally shaped by access to energy, rather than water.

Shelter from the physical environment

The *need* we have for shelter stems from our vulnerability to the outside environment and is directly linked to our lack of fur and our erect posture. Modern humans, like our closest primate relatives, have a sparse covering of hair that is much less effective at retaining body heat than the fur that covers the bodies of most other mammals. (Hair and fur are actually referring to the same structures. On fur-bearing mammals, some of the hairs are long and infrequent. They are called guard hairs and serve to keep the underlying fine hairs, called the downy layer, erect. If the downy layer is crushed, wetted, or oiled, its thermal protectiveness is lost. Humans have little or no downy layer.) Other warm-blooded animals that lack thermally effective fur coverings are elephants, rhinos, and hippos, plus naked mole-rats and bats. The former grouping maintains temperatures without the benefit of fur largely because of their sheer mass. Smaller mammals lacking fur must also seek refuge from temperature and moisture fluctuations that furry mammals ignore.[142] There is little reliable information on the degree to which human body coverings have changed over the past several million years, and we will not argue here that ancestral humans were as hairless as we are. However, paleoanthropologists strongly suspect that the lack of body coverings first appeared in early east African hominids 2 million years ago and probably reflects the tropical origins of our species. The subject of the timing and extent of hair loss, as well as the likely manifestation of body hair on Neanderthals and modern humans, has been considered recently by Berman (1999). She observes that the lack of fur must have placed considerable selection pressure for the manufacture of clothing during episodes of glaciation, but concrete evidence for clothing as a replacement for fur is only found in the late Pleistocene, approximately 40,000 years ago.[143] Clothing or animal skin wraps may have been used by earlier humans, but no evidence remains of these coverings. Without a doubt, our early human ancestors faced daily the daunting scenario presented at the beginning of this chapter.

Most paleoanthropologists believe that the drying and warming of the climate at the end of the Pliocene started a cascade of events that eventually led to the development of an upright posture in hominids.[144] An erect stature is advantageous in a hot climate because it minimizes the amount of body surface that absorbs direct radiation from the midday sun.[145] Excess heat is also released from the body by sweating, and this form of thermal protection is enhanced by hair loss because water can evaporate more freely from bare skin. However, all of the traits that help to dissipate excess heat during a hot day also result in increased heat loss during cold weather or at night. Exposed skin cools the body by evaporation (even at night), and an erect posture increases the exposure to wind that helps to dissipate heat much more rapidly even in the presence of body coverings. For naked, warm-blooded organisms, high wind speed is threatening at temperatures below 10°C and lethal below 0°C unless shelter from the wind is sought. Temperatures in this range do occur commonly in present-day savanna and most likely did as well at the time of early hominids.

This vulnerability to cold was most likely a distinguishing feature of the early hominids that emerged from their arboreal habitats in warm latitudes of Africa 1 to 2 million years ago. The open savanna, the area of which was rapidly expanding, was unlike the forests with their naturally reduced radiation load and wind speed. The first incursions into savanna quite likely would have required nightly retreat to the closed canopy of the forest for shelter. There clearly would have been an incentive to seek out locations that duplicated some of the benefits of a forest canopy as savanna began to replace forest in eastern Africa at the start of the Pleistocene and competition for the remaining forest habitat increased. Shelters at the bases of rock outcrops and cliffs would have been the only places[146] available in open savanna that would block radiation and wind and at the same time could accommodate modest numbers of hominids in social units that would persist over time. Individual acacia or baobab trees exist in savanna habitats, but such trees have little influence on wind speeds and would do little to protect sheltering humans from the cold at night.

Shelter from predation

As social colony–forming organisms of limited physical strength, early hominids would have represented a tempting food source for predators in Africa, Asia, and Europe. The fossil record shows no evidence that any morphological changes took place during our evolution that increased our ability to defend ourselves, unless one counts the evolution of brain power that increased our ability to avoid predator attacks. Their erect posture gave hominids such as *Australopithecus afarensis* a superior view of the landscape,[147] but the early detection of approaching predators was not sufficient to help them escape predation since most predators of grasslands and savannas hunted by stealth or in packs rather than by raw force or speed. Many small mammals avoid predation by digging burrows, but this is not feasible for larger organisms weighing more than 50 kilograms; such large dens in soil would be fundamentally unstable. Other primates seek safety in tree canopies, but the large-bodied hominids from *Homo habilis* onward have lost the ability to do so efficiently.

Among natural habitat features that could have offered our ancestors refuge from predators, rock shelters at the bases of cliffs stand out because of a number of characteristics that make them relatively safe and easy to defend. Caves are much more permanent than shelters formed of soil or wood. They can be attacked only from one side, and the talus slope provides a natural protective barrier: any approach to the cave entrance requires an upward traverse across a steep field of open rock, putting the attacker at a disadvantage. (The rocky debris that accumulates at the base of cliffs adopts a relatively predictable angle of repose between 34 and 40 degrees.) Considering these features, it is not really surprising that humans have been using rock shelters throughout history and up to recent times.[148]

Shelter from resource fluctuations

Our ancestors faced another serious problem as they began to exploit the savanna biome: the fact that resources such as food and water were no longer available year round. This is because the productivity of savanna is seasonally pulsed, unlike that of tropical forest, which is continuous throughout the year. The climate of savanna has a distinct seasonal cycle. The rainy season, during which most vegetation growth takes place, is followed by an extended dry season, during which much of the vegetation is dormant. In the evergreen tropical forest, in contrast, rainfall is abundant and distributed evenly throughout the year, resulting in constant productivity year-round. Various strategies to cope with this problem have evolved in the animal kingdom. Some mammals and birds display periodic hibernation or diapause, during which metabolic activity and consequently the need for food and water is vastly reduced. Such hibernation does not occur in primates, and it is therefore doubtful that it was ever used by early humans. Another strategy is the over-consumption of food or water when it is available, and storage of the energy derived from it in the form of body fat. Camels, bears, seals, and whales are the most familiar examples of organisms displaying such internal caching, but in primates the amount of food storage possible in body fat is limited. The two strategies most likely used by our human ancestors to cope with fluctuating resource supply are behavioural, rather than physiological: food caching and migration. (We have already established in a previous section that water could not be cached by early humans, and we will therefore limit our discussion here to the caching of food.) We will argue below that both these behaviours derived unique benefits from the presence of cliffs in the context of savanna.

Food caching is extremely common in vertebrates from shrews to large carnivorous felines to primates.[149] Sometimes the amount of food in the cache is sufficient for only a few days, as is the case with carcasses stored in trees by large cats. Other species such as beavers or muskrats store enough food to supply adults and young for weeks or even months. Seeds, buds, twigs, tubers, and other plant foods are commonly cached underground. Rodents including rats, mice, voles, and shrews may build middens that store large volumes of material. Such caches are subject to competition or predation from other organisms, and are often defended vigorously. Seeds buried individually (such as acorns by gray squirrels) attract fewer competitors, but caches are not marked and therefore relocation success is often low. This has selected for the behaviour of massive food overburial to increase the likelihood of food retrieval from at least some of the caches.

The caching of meat presents entirely different challenges. Animal carcasses undergo significant bacterial decomposition within 48 hours of death,[150] in particular when in contact with the ground or with water. They also attract carrion feeders and scavengers, many of which are able to fly or climb trees, making the caching of meat in trees ineffective.

Early humans would have had to protect their food caches from a wide variety of predators and scavengers. These would have included members of the dog family (wolves, foxes, jackals, and hyenas) and the cat family (lions, leopards, jaguars, and tigers), bears, peccaries and pigs, mice and rats, as well as scavenging birds such as rock doves and vultures. Many of these species were much more physically robust than humans, resulting in strong selection pressures on humans to develop food caching techniques that the competitors could not take advantage of by using brute strength alone. Food caches would have also needed to be defended from other humans; in fact, competition for resources is usually greatest among members of the same species because the resources they need are the same.[151]

For all these reasons, the location of the food cache would have been a crucial factor in determining the success of the caching strategy. The best locations for early hominids to cache resources would have satisfied the following five basic requirements: (1) the resource would not escape or deteriorate while being moved to the cache, (2) the resource would not escape or be lost from the cache, (3) other organisms could not find the cache and consume or contaminate the resources, (4) the cache could be defended from attack, and (5) the cache could be relocated when needed. There are few places in open savanna where these conditions would have been satisfied, but cliffs and the caves within them would have been uniquely suited (Figure 19). As already pointed out, caves are easy to defend and are relatively inaccessible to predators and competitors. They are also easy to find for retrieval of the resource.

The cool and dry conditions within the caves would have provided optimal conditions for long-term food storage. An additional benefit most likely arose after *Homo erectus* acquired the use of fire. Smoke from fires close to the cave mouth would have reduced parasite loads[152] and kept away scavengers. And the smoke-clouded vestibules in caves may have represented one of the few settings that would accidentally result in the curing of meat and fish. This would have increased the safety of later consumption given the sensitivity of the human gut to bacterially contaminated meat.[153]

There is evidence for a second strategy used by early humans to avoid periods of low resource availability: migration. Expansion to new, previously unexplored terrain can open up huge opportunities but also carries large risks due to the colonists' lack of knowledge of the habitat and the organisms already occupying it. In contrast, cyclical movements coupled to seasonal changes in temperature or precipitation carry the benefits of intimate familiarity with the habitat. For this benefit to be realized, however, the familiar landscape must be recognizable from a distance as the place to return to. Cliffs, unlike any other naturally occurring structure, would have represented permanent and highly visible landmarks for orientation within an otherwise featureless savanna and could have been described to others even before language development took place.[154]

Tools

Humans are primarily distinguished from non-human primates based on their tool use and the development of cultures associated with those tools.[155] Tools are defined as physical objects used to perform specific tasks that an organism cannot complete with its own body. By this definition, the rocks used by sea otters to open sea urchins and the straws used by chimpanzees to harvest termites are tools, but not the silk of spiders or the traps constructed by venus fly trap plants.[156] The use of tools can be learned by imitation without being understood by the user.

The principal distinction between the tools used by people and those used by other organisms is that human tools are crafted from naturally occurring materials, while other organisms use the raw material "as is." For example, stone chopping tools recently estimated to be 2.5 million years old and attributed to *Australopithecus garhi*[157] were routinely, carefully, and intentionally modified for specific purposes, including the harvesting of food and possibly protection from predators. Estimates of the initial date of colonization of the Americas are based on the earliest evidence of manufactured tools. At present, colonization is thought to have taken place about 11,000 years ago and tool-like artifacts that have earlier dates are considered to be of natural origin. However, debate currently

Figure 20.
The latest and the largest stone tools ever manufactured included gritstones such as these carved directly from sandstone outcrops at Froggatt Edge, Derbyshire, U.K. Stone tools extracted from rocky outcrops and escarpments have been manufactured for at least one million years. Photo: D. Larson.

centres on the question whether 30,000-year-old rock fragments found at the base of a tall cliff in Brazil were formed by natural processes or represent tools used by people.[158]

All of this suggests that tools were among our real needs right from the time when we became human. The importance of tools went beyond the procurement and processing of food; Bingham (2000) has argued that most of human evolution and history can be explained on the basis of how humans have learned to use tools to kill or punish opponents from a distance. The earliest tools used by hominids were rocks excavated from cliffs,[159] and only later were bone and wood used in tool manufacturing. Stone tools such as the millstones pictured in Figure 20 have continued to be in use until modern times. Materials used to fashion stone tools include chert, flint, obsidian, schist, quartz, quartzite, jasper, and other fine-grained rocks. Only volcanic or thickly bedded sedimentary rocks were used, since granite and thinly bedded sedimentary rocks like shale are unsuitable for tool manufacture. The only places within the savanna biome where such materials would have been easily accessible to early hominids and available in sufficient quantity were probably cliffs and rock outcrops. Given the importance of tools to human evolution, one could claim that the choice of habitation sites that could at the same time supply materials for use in tool manufacture would be strongly selected for.

Knowledge and spiritual resources

Humans as we now know them have a fundamental need for knowledge based on reasoning rather than simple rote learning. Knowledge acquisition in its simplest form, the acquisition of learned responses to environmental stimuli, is evident in all vertebrates and even in some invertebrates such as insects and snails. Knowledge obtained by reasoning is only evident in humans, although some argue that primitive reasoning may be found in some primates and in whales and dolphins.[160]

If knowledge based on reasoning was transmitted between humans early on in our evolution, it must have been by means that were impermanent such as oral and gestural communications or temporary inscriptions. The earliest field evidence of such transmission appears about 40,000 years ago in the form of sculpture, inscriptions, and paintings, all of them associated with the remains of modern humans found in rock shelters. The earliest evidence of music, in the form of some primitive flutes, is also found in such sites.

The subject of much of this early transmission of knowledge is thought to be related to another basic human need: spiritual resources. Boyer (2000) argues convincingly that belief systems, including the development of cults and religious systems, are products of natural selection for social interactions that bind groups of hominids together. He argues further that the intangibility of the spiritual world actually reinforces the belief systems since they cannot be tested or challenged using empirical evidence. Whatever the belief system, humans have for at least 40 millennia recorded different aspects of their spiritual lives on surfaces that resist decomposition: rock walls, cliff faces, cave walls and ceilings, and individual stones. Much popular attention has been directed to the oldest evidence of spiritual awareness that is present in the caves of southern Spain and France, but rock art is plentiful and spans the period from 25,000 years ago to the present.[161] Spiritual objects may have been recorded on impermanent media as well, but obviously we have no evidence of this.

Current demographic surveys suggest that most people today believe in some form of supernatural control of their lives and the universe.[162] Burial practices, objects of art, forms of artistic expression, and numerous other sources of evidence all suggest that humans need to express a belief that there are both benevolent and hostile forces in nature over which they have no control. Lack of these resources may not be fatal, but there appears to be good evidence that psychological illness of individuals and the population as a whole can occur when no spiritual resources are obtained during life.[163]

In this chapter we have marshalled evidence to conclude that cliffs and rock shelters within the savanna biome were optimal places for humans to exploit. They represented an easily recognizable base from which to forage for food and water, offered shelter from the physical environment and from predators, provided safe and permanent places to cache food, were a source of materials to make tools, and were the only available media for the permanent recording of spiritually important images. In the previous chapter we showed that a wide array of opportunistic plants and animals also made these rock outcrops their home, and that many of them were edible or in other ways useful to humans. This combination of "rock as home," "rock as sanctuary," and "rock as supermarket" is, we argue, compelling. No other combination of habitat elements, in our view, has supplied the resources necessary to sustain human life, and the fossil and archeological record supports this conclusion.

In the next chapter, we will show that the exodus from the rock shelter environment never really happened, or that it happened in a way very different from how most people understand it. We will show that humans began to modify their natural rock shelters progressively, eventually becoming capable of constructing such shelters themselves. At this point, the inherent portability of the built environment freed people to exploit sources of energy and water wherever they occurred. At the same time, the newly constructed built environment also provided habitat for all the plants and animals that had been our companions in the rocks.

---

90   See many recent volumes including Shreeve (1995), Gamble (1986, 1994), and Tattersall (1998).
91   We will use the name *Homo neandertalensis* instead of *Homo sapiens neandertalensis* because of recent evidence rejecting the idea of hybrid formation between Neanderthals and modern humans.
92   Kunzig (1999).
93   Stringer and McKie (1996), Isaac (1975), Butzer and Isaac (1975).
94   Larick and Ciochon (1996).
95   Swisher et al. (2000).
96   Gamble (1999a).
97   Tattersall (1995).
98   Shreeve (1995), Gamble (1994).
99   Sherratt (1980).
100 Stringer and McKie (1996).
101 Reed (1977a, b).
102 Vrba et al. (1995).
103 Tattersall (1998).
104 Orians (1986), Larick and Ciochon (1996).
105 Leakey (1960).
106 Bar-Yosef (1995).
107 Larson et al. (2000a).
108 McGrew et al. (2003).
109 Pfeiffer (1972).
110 De Lumley (1975).
111 Schepartz et al. (2002), Binford and Ho (1985).
112 De Lumley (1969).
113 Examples are: Le Moustier, France; Monte Circeo, Italy; Uzbek, Central Asia; and Shanidar Cave, Zagros Mountains, Iraq (Solecki 1963); les Eyzies, France, and Saint-Césaire, France (Lévêque 1993; Patou-Mathis 1993); Combe Grenal, France; Lacave, France; and Mount Carmel, Israel (Smith et al. 1999); Gorham and Vanguard Cave, Gibraltar (Gamble 1999b); Abri Romaní, El Juyo, Spain (Carbonell and Vaquero 1998; Pokines 2000); Mezmaiskaya Cave, Russia (Harvati and Delson 1999); Black Cave, Turkey (Otte et al. 1998); Apidima Sea-Caves (Harvati and Delson 1999); Sally Binford's 1962 site on the sea of Galilee, and the Pilanduk and Tabon Caves that were occupied in the Philippines for over 40,000 years (Fox 1978).
114 Smith (1992).
115 Smith (1992).
116 Caldwell (1977).
117 Buckler et al. (1998); Wright (1977), and MacNeish (1964, 1977); and Kraybill (1977).
118 Rudofsky (1977).
119 Deacon (1975), Clark (1975), Gamble (1999a).
110 Bordes (1972).
121 Gargett (1999).

122 Kurtén (1976).
123 Larick and Ciochon (1996).
124 Potts (1984).
125 White and Brown (1973).
126 Krings et al. (1997, 1999, 2000), Ovchinnikov et al. (2000).
127 Wright (1977), Reed (1977a, b).
128 Howell (1965), White and Brown (1973).
129 1 KJ (kilojoule) = 0.2389 Kcal (Kilocalories).
130 United States Food and Drug Administration (2003): website www.fda.gov/cder/
131 The zone of thermal neutrality is a narrow range of environmental temperature within which a warm-blooded animal needs to expend no metabolic energy in order to maintain its body temperature. At temperatures outside this zone, the organism needs to expend metabolic energy either to keep itself warm or to cool itself down. For humans, the zone of thermal neutrality is approximately 15–25 ℃; the actual temperature span varies among individuals as well as among species.
132 Pure carbohydrates yield 18 KJ of energy per gram of dry weight, and fats yield double this value. Unpurified plant foods, however, include 30–50% of components such as cellulose fibre that are indigestible to humans, thereby lowering the effective energy content to approximately 10.5 KJ per gram.
133 Tattersall (1998).
134 Milton (1999). The question of the degree to which humans might have acted as scavengers vs. hunters may never be perfectly answered. It is probable that our feeding activity was controlled by local or regional circumstances that cannot be generalized.
135 Ragir et al. (2000).
136 Ragir et al. (2000), Marean (1998), Villa and Soressi (2000).
137 Harper (1977).
138 Harper (1977).
139 United States Food and Drug Administration, 2003: www.fda.gov/cder/
140 250 mm of water over an area of one square meter = 25cm x 100cm x 100cm = 250,000 cm^3 = 250 litres.
141 500 mm of rain (per year) over an area of 1 m^2 = 50cm x 100cm x 100cm = 500,000 cm^3 = 500 litres per year; to supply the requirement of 1460 litres per year, an area of 1460/500 = 2.9 m^2 is needed.
142 Schmidt-Nielsen (1975), Monteith and Unsworth (1990).
143 Johanson and Edgar (1996).
144 Shreeve (1995).
145 Gibson and Nobel (1986). While this book is about cacti, the principles of how radiation load is minimized by the columnar shape of cacti apply just as well to humans.
146 Larson et al. (2000a).
147 Tattersall (1998).
148 Mercader et al. (2001).
149 Tattersall (1998).
150 Ragir et al. (2000).
151 Keddy (1989).
152 Rudofsky (1977).
153 Ragir et al. (2000).
154 Personal correspondence with Dr. L.Schepartz after the appearance of an article in *Earth* magazine about her discoveries of large mammal butchering sites in China (1996).
155 Diamond (1999), Sherratt (1980).
156 Van Schaik et al. (1999).
157 Johanson and Edgar (1996).
158 Stringer and McKie (1996).
159 Van Schaik et al.(1999).
160 Tattersall (1998).
161 Jerardino and Swanepoel (1999), Deacon (1999), Stoffle et al. (2000), Tilley et al. (2000), Lewis-Williams (2001).
162 Tilley and Bennett (2001).
163 Boyer (2000).

CHAPTER FOUR

From Caves to Cathedrals: The First Exodus

The Urban Cliff Hypothesis is about an exodus that could be called the *first exodus*: the movement of *Homo sapiens* from caves to buildings then on to towns and cities. We argue in this book that the first exodus is nothing other than a movement from one kind of rock shelter to another kind of rock shelter: from the natural rock shelter to the built rock shelter.

The common perception of how civilization arose goes like this. At first, we were wild intelligent primates living in caves. Then we developed architecture, moved to environments we built ourselves, and started to populate towns and cities.

This view jumps over the fascinating link between caves and architecture. As pointed out earlier, we often associate cave dwellings with negative images of people grunting out a hand-to-mouth existence: the derogatory term *troglodyte* means nothing other than cave dweller. People living in caves have been given a bad rap by Western journalists. For example, the negative portrayal of terrorists in Afghanistan is reinforced by references to their cave-dwelling behaviour. Yet, as we showed in the previous chapter, caves and rock shelters are in many ways ideal places that provide for many of our basic needs. Many peoples have lived remarkably full lives in cave dwellings throughout our history and into the modern era. Cliffs and caves, we believe, were appreciated by their residents not only for their functional advantages, but also as places of spectacular beauty.

If all of this is true, then why did the first exodus happen? In this chapter we will consider the drawbacks of natural cave dwellings that might have prompted our ancestors to modify their rock shelters in an attempt to make them more useful and comfortable. At some stage in our evolution, this became possible as we developed the capacity to alter the physical structure of rock shelters and caves. The alterations later involved the movement of individual rocks, their use in the construction of buildings, and eventually the manufacture of the building materials themselves. We will show that the earliest built structures closely resembled naturally occurring dwelling sites, and demonstrate that this similarity allowed other cliff-dwelling organisms to take over built structures as their habitat. Finally, we will evaluate the degree to which this resemblance applies to modern built structures, and argue that most of us are still cave dwellers even though some of the caves look rather fancy and cost thousands of dollars per square meter to rent or own.

Problems with cave living

The act of building shelters is not unique to humans. Birds build nests, badgers and muskrats dig burrows in banks, bees construct hives, and many other organisms build protective structures for sleeping, breeding, or escape from competitors or predators.[164] However, humans are descended from organisms that do not allocate significant effort to the building of shelters. Modern species of chimpanzee, gorilla, and bonobo find safety from predators in the forest canopy and may gather branches and leaves for protection from the weather. They do not actively construct permanently occupied shelters,[165] although they do sometimes exploit naturally occurring rock shelters.[166]

When *Australopithecus* species began to adopt non-arboreal lifestyles in savanna environments in east Africa and lost the ability to use trees, they sought shelter in caves. The reasons why caves were the refuge of choice have already been explored in Chapter 3. The important point here is that, for more than a million years, caves were used as dwellings without modification. As far as can be told from the Pleistocene fossil record, early humans did nothing to protect themselves from heat or cold except to use rock shelters and fire. In many ways, these early humans used cliffs for shelter in much the same way as do present-day baboons.[167]

Natural rock shelters and caves were ideal campsites in the sense that they supplied many of the basic needs of early humans, but they supplied little in the way of comfort. As humans began to conceive of *wants* in addition to *needs*, natural caves left much to be desired. As we will show below, some of these problems were relatively easy to address by simple modifications of the caves themselves. Others were not, and these are the ones that ultimately provided the incentive for the exodus to the built environment.

The regulation of microclimatic factors such as wind, temperature, humidity, and light is now known to be centrally important to the maintenance of human comfort in buildings.[168] Most natural caves would have provided some degree of basic protection from the elements, and temperatures in the interior of large caves are known to be remarkably moderate and stable year-round. However, an unheated cave entrance would not have been especially comfortable, and humans probably began to use stones or animal skins to block out rain or wind from the cave opening well before the formal built environment appeared. The use of fire that began with *Homo erectus*[169] would have begun to make caves more habitable, but accumulations of smoke would have represented a hazard. This hazard most likely explains the placement of most hearth sites at the mouths of caves just slightly behind the drip-lines of the cliffs.[170] In these locations, some warmth would radiate to the cave interior, but toxic fumes would dissipate to the outside. Other sources of discomfort in caves that could have been improved with minor modifications include uneven rocky floors and ceilings formed of loose and dangerous rock slabs. Also, most caves would have naturally provided places to store food or tools, and slight modifications would hugely increase the effectiveness of the storage sites and caches.

Although caves are more easily defended from predators and scavengers than campsites in open savanna, there are no structures to protect a cave that is left unattended. Food caches, waste associated with hunting,[171] and human organic waste would have attracted to rock shelters an abundance of scavenging vertebrates such as hyenas and wolves. This made it necessary for a fraction of the human population to remain behind and defend the cave when other members of the group were foraging for food and water. The construction of walls or barricades would have been a simple step to significantly improve the security of the cave and make it easier to defend.

Another drawback of rock shelters and caves was their fixed location. The surrounding cliffs were home to a wide variety of edible plants and animals, but the productivity of rocky habitats is too low to sustain significant numbers of humans (as we have shown in Chapter 3). For this reason, all easily accessible resources were quickly depleted, and foraging had to take place at a distance from the shelter site. In most cases, water had to be obtained away from the cave opening, as well, since reliable sources of water within caves are only common in active karst areas.[171] After active foraging in outlying areas, the humans returned to the cave site where the game was processed, tools were fabricated, and individuals could rest. The important point is that before the construction of shelters, each habitat element had to be accepted and exploited optimally as it stood. For example, Binford (1984) studied a well-known site at the Klasies River mouth, South Africa, which was occupied between 125,000 and 50,000 years ago. He showed that in the latter part of this period, the construction of shelters and the ability to control fire made it possible for people to concentrate all their activities in one place. People could secure resources, prepare food, and sleep in one location. The ability to build things meant that people were less enslaved by nature: they could reshape nature to suit their need. Building represented a major ecological breakthrough because dwellings could be constructed nearer to resources. It also permitted the development of a more intense division of labour between the sexes and among the age classes.

Finally, the strongest incentive for the first exodus was probably population growth and lack of space. There is a finite supply of habitable caves and rock shelters in any given area. In Chapter 2, we showed that the total area of cliffs in relation to other habitat area is vanishingly small, rarely exceeding 1%. Moreover, not all of these cliffs will contain caves. The extent to which they do depends on the type of rock that forms the cliffs: of the 114 areas in the world with well-developed caves, 80 are formed of limestone, 10 of sandstone, and 9 in lava tubes, but only 15 others consist of conglomerate, gneiss, schist, granite, and basalt.[173] Of the many different types of rock that can form cliffs, therefore, extensive caves are likely to occur with any frequency only in sedimentary rock such as limestone and sandstone. Even for rock types that do form caves, the frequency of caves large enough to be useful as a shelter for a family group is small. To cite again the example used in Chapter 2, for the 350 hectares of Niagara Escarpment cliffs in southern Ontario, there are probably fewer than 20 such caves.

Each cave itself has a limited capacity to house humans and cannot easily be enlarged. As a result, the availability of living space must have severely limited the postglacial expansion of the human population. Smith (1992) estimates that immediate postglacial Britain had, at any one time, a total population of only 250 persons, while France, where glaciation was less severe and the occurrence of caves was greater, was home to perhaps 10,000 to 15,000. Since cave dwellings cannot be easily adapted to house

FROM CAVES TO CATHEDRALS

more people, there would have been a strong incentive to migrate to new territory or to develop new ways of accommodating human population growth. The construction of buildings not only would have provided ample space, it also would have allowed for the spatial separation of functions such as sleeping, food preparation, living space, waste elimination, and the interment of the dead.

From caves to built structures

The exodus of humans from caves to built environments arose from the desire to eliminate the disadvantages of caves and rock shelters while at the same time maintaining all their advantages. There is more than one way to achieve this goal, and different peoples throughout history have used different approaches, depending on the technology available to them.

The simplest technique is to modify existing caves, and Kostof (1985) argues that caves can take on all the greatest benefits of a built structure with marginal modification. Carving artificial caves into cliffs requires more advanced technology, but this will vary depending on the type of rock forming the cliffs. Freestanding structures away from cliffs can be built of easily available materials such as rocks and bones, but such buildings tend to be temporary. Permanent structures are made possible by more advanced materials that are themselves manufactured, such as wooden beams and bricks. These different building techniques do not appear sequentially in the historical record, but appear at different times in different parts of the world and often co-occur. Even today, many primitive tribes in northern Africa and western Asia occupy a continuum of dwellings ranging from natural caves to man-altered caves to mud-brick huts that resemble caves.[174] These different types of habitations are often physically adjacent, as for example in the case of Dogon villages in Mali.[175]

The idea of manufacturing a building to replace the functions provided by a cave was probably deeply rooted in the culture of the humans of the time. Furthermore, just as cultural variation occurs in societies today, some cultures were probably more adept at dwelling construction and exploited the opportunities provided by it, while other cultures shunned these developments.[176]

Simple modifications to cave entrances, such as the construction of windbreaks or short retaining walls in the front section of the dwelling site, first appeared as early as 70,000 to 65,000 years ago, for example at sites in southern France that have continued to be occupied until modern times.[177]

Figure 21.
In Cappadocia, Turkey, cave dwellings have existed for more than 1000 years and are now an important part of the local economy, as illustrated by the Gamirasu Cave Hotel. Brochure courtesy of Süleyman Çakir, proprietor.

In the Cappadocia region of Turkey, habitable shelters that have been carved into volcanic and conglomerate outcrops (Figure 21) originated in historic times, but there is evidence that alteration of the relatively soft rock in this region to create human dwellings goes back to the Paleolithic between 60,000 and 40,000 years ago.[178] Similar modified caves and dwellings carved into cliffs are found (and often are still occupied) all over the world, but most of them are thought to have origins dating back no more than 8000 years. These include cliff dwellings at the base of high mountain peaks in southern Tunisia,[179] in France (Figures 22, 24), in Ireland,[180] and elsewhere in Europe (Figure 23).[181] In Arizona, the Pueblo Indians built entire cities under the protective edge of a large rock escarpment.[182]

Evidence shows that during the period from 70,000 to 40,000 years ago, the human modification of cave sites increased and there was some purposeful construction of other kinds of temporary or more permanent shelters. Gamble (1999a) and others claim that clusters of large rocks dating from more than 300,000 years ago represent the supporting structures of dwellings built much earlier by Neanderthals, but there are no organic

Figure 22.
At Les Mèes, France, conglomerate rock that forms vertical pinnacles has
been used to form carved dwellings for people and livestock for centuries.
Photo: D. Larson.

S. B.-G.

HABICHSTEIN, BOHEMIA

A castle belonging to Count Wallenstein, now abandoned owing to the falling away of portions of the rock. It contains stables for horses and cattle, now inaccessible without ladders.

Figure 23.
Occupied cliff dwellings in the Czech Republic. From Baring-Gould (1911).

CAVE DWELLERS AT DUCLAIR

These are typical of countless others on the Seine, the Loir, the Loire, and its tributaries, as also on the Dronne and Dordogne.

Figure 24.
Occupied cliff dwellings in France.
From Baring-Gould (1911).

remains associated with these open-air sites to support this interpretation. Even if this early evidence is real, there is no indication that any other built structures were in existence until the Mousterian period, between 60,000 and 40,000 years ago. At that time, convincing evidence for the active construction of buildings begins to appear in the archeological record,[183] and between 40,000 to 20,000 years ago there is an explosive increase in the occurrence of such structures. The first built structures were simple, non-permanent compositions of pre-formed rocks, mammoth skeletons, or skins. Circular concentrations of mammoth bones in Moldova have been interpreted as dwellings constructed during the middle Paleolithic, more than 50,000 years ago.[184] Stone tools, animal remains, and fire pits have been excavated from within these bone circles, suggesting that they may have represented relatively permanent encampments. Similar circular structures made of stone represent the first buildings constructed by the Natufians 15,000 to 12,000 years ago in an area that is now Israel and Palestine.[181] There, the transition from cave dwellings to these simple rock dwellings occurred during a prolonged period of climatic cooling and drying and coincided with the first appearance of primitive agriculture.

In his book *Timewalkers* (1994), Gamble suggests several reasons for why the building of dwellings is such an amazingly recent development in human history. He argues that building was made possible not only by developments in engineering and tools that had not been available before this time period, but also by an expanding awareness of the power of collective intelligence present in the society as a whole. The rapid increase in human population size that coincided with the appearance of built structures is explained by Shreeve (1995) in this way:

> I can think of only one resource on earth with so much raw value and dangerous potency that people had to invent art and style and perhaps even new forms of language and consciousness, just to keep it from blowing everything apart. Other people. . . . What was truly revolutionary about the late Paleolithic was not language, style, or art, but the opening of social conduits through which information in all such novel forms could flow. . . . It does not take a new or special kind of brain to make a bone harpoon point, or to learn how to use it. But it does take a chain of brains to keep the idea alive. It was the human need to reach out for other humans, across landscape, that fueled the first signs of the creative explosion.

Shreeve argues further that one of the main factors that spelled doom for Neanderthals was that, as a species, they were conservative in their culture, considering change anathema. Hence, they would have resisted any opportunities to communicate with neighbouring cultures and share knowledge-based resources such as new building techniques. In contrast, early *Homo sapiens* embraced, or at least readily exploited, the culture of neighbours. This ready acceptance of change might easily explain the rapid transition from cave dwelling to built dwellings about 40,000 years ago. Interestingly, Gamble (1994) describes a wide variety of other cultural changes taking place at about the same time. These include the first expressions of graphic images or art that probably served to provide people with education and spiritual foundations.

Even though huts were in sporadic use by 40,000 years ago, evidence of the first permanent villages dates only to the Holocene. Permanent rectangular buildings constructed from mud or stone bricks appeared around 10,000 years ago at Jericho, Çatal Hüyük, and Jarmo.[186] In all of these cases, the village sites were permanently occupied over several generations and protected by large fortified stone walls, guarded, in some cases, by towers.[187] From these earliest examples of permanent and engineered dwellings to the present, human social development and cultural evolution has centred on the permanently occupied village.

Characteristics of buildings

The first human dwellings
The earliest built dwellings copied many of the most attractive features provided by caves while also offering a number of improvements. All of the first brick structures were extremely solidly built and evidently meant for more than temporary use. Their thick outside walls guaranteed a physical buffer from the elements; Legge (1972) argues that the conditions within the large stone houses built in Arab countries today are similar to cave microclimates, and this was probably true for the first built dwellings as well. Entries and exits were limited to narrow doorways or in some cases occurred through the roof of the huts. (The latter type of entry was used, for example, at Çatal Hüyük in Turkey.) This provided protection from predators and restricted access to the food caches by competitors.[188] Since places to cache water and food were designed into the structure, they could be manufactured to resist predators both large and small. The dwellings could be built to have moderately flat and comfortable working and sleeping surfaces. Primates are the only vertebrates to sleep on their backs, and the construction of a surface sufficiently flat for sleeping comfortably represented a huge improvement over the uneven floor surfaces of caves.

The greatest benefits of built structures, however, arose from walls and their ability to segregate different life functions that could then be carried out with greater efficiency. Even in caves, activities such as sleeping, tool manufacture, the butchering of game, and the creation of art were assigned to different locations. Walls, however, allow people to communicate with each other when they wish but also to remain private when they need to. Without other activities taking place simultaneously, the effectiveness of knowledge transfer is enhanced and the opportunity for intellectual development is increased. The design and architecture of living spaces will therefore directly influence the exchange of ideas, of information, and therefore of culture as a whole. Early cave dwellings in Scotland had only one wall separating the human living space from that of the animals.[189]

Other early buildings had at least three subdivisions of space: dwelling and sleeping spaces, food preparation areas, and storage chambers for tools and food.[190]

The technology to manufacture stone or mud bricks opened up new opportunities to humans.[191] Archeologists have focused on the advances made possible by different configurations of stone tools, but only rarely mention the enormous advancement made by the brick. The brick as a tool was entirely responsible for allowing humans to escape from cave dwellings while retaining all of the benefits that caves provided. Wood beams undoubtedly come a close second as the most important building materials ever invented. When beams were finally incorporated into the design of the structures, roofs could be made safe from collapse, and doors could be constructed to completely exclude large competitors and predators. Protective walls resembling cliff faces could be built around the dwellings, forming a second layer of defence against both the physical elements and human or animal attackers.

The buildings themselves, as in the case of both Jericho and Jarmo, could be constructed in a location that provided an abundant supply of water. The physical dimensions of the huts could be adjusted at any time to accommodate large human populations. Nothing about the building materials limited the number of persons who could be accommodated. The first dwellings were built in a time period in which most food was obtained by foraging rather than by cultivation, but Thorpe (1996) and others argue that climate change was the stimulus for both the development of primitive agriculture and the first appearance of built structures. He argues that at a time of climatic cooling, reliance on a cave as a base camp for hunting and gathering would not have allowed people to take advantage of the productive potential of sites near stable sources of water. Human survival would have been greatly increased by moving seeds near such sources of water and by constructing buildings close by to allow the sown seeds to be tended. Once villages containing continuously occupied structures were built, there was strong selection for the incorporation of plant and animal species into cultivation or husbandry. The species that were most likely to be incorporated were species that were already familiar to humans, could easily exploit the environments surrounding the human dwellings, and offered a substantial return on the energy invested in cultivation or husbandry.

Modern built structures

Despite the various configurations of external architectural design, the basic living units within which people sleep, work, and play have changed little between the Neolithic and modern times.[192] People still place great value on features of built structures that provide security, energy, water, sanitation, privacy, chances to mix with other people, and opportunities for other plants and animals to live. Solidly built structures with ceilings, walls, windows, basements, kitchens, and workshops are present both in the villages of the !Kung bushmen of the Kalahari and in the city of Toronto. All of this suggests that ever since the first human-built structures appeared, the development of architecture has served the goal of designing a better cave dwelling.

From the Neolithic to the Middle Ages, walled structures surrounding castles, fortresses, villages, towns, and even cities were exceptionally common. In almost all of these structures, massive stone walls around the perimeter enclosed an inside space that was subdivided for separate activities such as sleeping, eating, tool construction, food storage, ceremony, waste disposal, and recreation. The main function of the external wall was to keep out enemies, but it also provided some protection from a harsh climate. Defensive structures were commonly built into the walls of fortresses, castles, and even early Christian churches and cathedrals.[193] In most cases, the walled structures were themselves built on heights of land, often directly on the pediments of existing escarpments or cliffs[194] (Figure 25). Such fortresses were often described as cliff-like on the outside and were almost impossible to conquer without a siege.[195] Examples of such fortresses constructed on existing rock structures are found throughout Europe, Asia, and Africa, including Greece[196] (Mycenae and Meteora), Great Britain (Edinburgh and Nottingham) (Figure 26), Jerusalem, Cambodia,[197] Sri Lanka and Zimbabwe (which translates to "house of rock"). An attacking enemy could be much more easily handled from a height of land above the attack point, and the use of both castle keeps and cliff edges for defensive positions is well established.[198] Despite the costs associated with building such walled structures, the positive benefits have been massive. Present-day architecture has retained, at the scale of an individual building, many of the features present in early walled towns and fortresses.[199]

The structure of modern cities and towns is similar throughout the world, with solidly built tall buildings packed closely together in the central core (Figures 27, 28). These "concrete canyons" are not merely

Figure 25.
In the town of Sisteron, in the south of France, modern stone buildings have been built along the base of natural escarpments that in previous centuries have been used to create the pediment for even larger human-built escarpments in the form of protective fortresses. The co-occurrence of these structures in this image lies at the foundation of the Urban Cliff Hypothesis. Photo by D. Larson.

similar to cliffs containing large numbers of cave dwellings; to a very real extent, they *are* cliff and cave environments. The subdivided living and working components within these buildings are little more than modern versions of the basic living and working spaces that originated in caves and subsequently emerged in primitive architecture.[200] Western city dwellers may speak disparagingly about living and working in these concrete

Figure 26.
Rock-cut dwelling in sandstone at Nottingham, U.K. The natural protection
and permanence provided by the rock resulted in the construction of these
cave dwellings and also the main castle buildings of the city.
Photo: D. Larson.

canyons, but a substantial economic value is placed on the occupancy of
vantage points: apartments or offices with a view command a higher price
than those whose external sight lines are confined by adjacent buildings.
Penthouse apartments are nothing other than cave dwellings located near
the top of the rock outcrop, occupying a position that affords them the
greatest view over the landscape (Figure 29).

Most contemporary buildings have a spatial configuration that allows
natural light to penetrate, provides locations where indoor plants can
prosper, and includes small alcoves where people can rest quietly. Aspects
of a natural environment incorporated by many modern building
designers include rooftop gardens, atria, streams, living walls, and tree-
lined walkways.[201] Such features function to sustain the air and water
quality in the building itself.[202] The Canada Life building in downtown
Toronto includes living plants and microorganisms in a rock-wall setting
that effectively provides some filtering of the air and water resources of the
building. Large business towers in cities recreate the cliff environment to
the point of providing breeding sites for the once-endangered cliff species,

Figure 27.
Modern urban cliff environment in the central core of Toronto, Ontario.
In this instance, sculptures of cattle resting in an open grassy savanna reflect
more reality into the scene than the artists and architects intended.
Photo: P. Kelly.

103

Figure 28.
The urban cores of cities such as New York City include street scenes that
recapitulate natural escarpment scenes around the world. Photo: P. Kelly

Figure 29.
A North Carolina Department of Tourism brochure that emphasizes the psychological connection between living on the edge of a lush escarpment (in this case, Linville Gorge) and living in a penthouse apartment. Photo courtesy of North Carolina Department of Tourism. Used with permission.

the Peregrine falcon (*Falco peregrinus*).[203] Table 5 compares many of the criteria that allow both caves and buildings to provide suitable human habitat.

According to O'Sullivan (1994), modern humans all over the world spend roughly 90% of their time confined to buildings. In the next chapter, we will try to understand why humans are emotionally comforted by built structures. We will argue that people actively seek out these kinds of locations to live and work because in part they reflect the original nurturing environment of natural rock shelters. The statement "my home is my castle" expresses the feeling associated with complete authority over one rock outcrop from which intruders are excluded.

Walls of opportunity

As they attempted to duplicate the benefits of rock shelters, humans ended up creating environments that closely resembled natural cliffs, rock outcrops, and caves. This opened up huge opportunities for the organisms

TABLE 5

Fifteen criteria used by Samuels and Prasad (1994) to evaluate the degree to which a building achieves the goal of producing a suitable human habitat. To this list of criteria we have added our own comments about how each criterion is met both within the built landscape and in the natural landscape of cliffs and caves.

CRITERION	BUILT LANDSCAPE	NATURAL CLIFF AND CAVE LANDSCAPE
1. Improvement of air quality	low emission of combustion products by efficient fuel systems and ductwork	hearths near cave mouths behind drip lines
2. Improvement of water quality	low effluent load and centralized filter systems	latrines distant from water sources
3. Use of rainwater	rooftop collection systems	cliff/talus biota dependent on runoff
4. Production of own food	rooftop and balcony gardens	plateau, ledges, talus, and slopes all used as sources of plant/animal material for gardens
5. Creation of productive soil	internal building wet recycling programs	latrines used for growth of camp-following species
6. Use of solar energy	orientation and shape of building used to maximize solar radiation in winter and minimize it in summer	cliffs/caves selected partly on the basis orientation/slope
7. Storage of solar energy	mass of building and internal heat sinks used to store radiant load	rock outcrops act as heat/cold sinks
8. Sound absorption	insulating materials used to isolate subgroups of humans	rock walls impenetrable to sound
9. Recycling of own waste	internal sewage recycling systems	latrines used to recycle waste into edible commodities
10. Self-maintenance	minimum requirement for external energy inputs	some cliff cave sites occupied for 40,000 years with no supplementary energy inputs
11. Matching of nature's pace	building adds a sense of permanence and tranquility	occupied because of permanence and proximity to resources
12. Provision of wildlife habitat	all buildings do this without many people being aware	all cliffs and caves do this naturally
13. Provision of human habitat	buildings specifically constructed for this	caves specifically selected for this
14. Moderation of climate	orientation and design of building used to maximize solar radiation in winter and minimize it in summer	cliffs/caves selected partly on the basis of orientation and exposure
15. Creation of beauty	by copying the natural structure	n/a

that had shared cliff habitats with early humans and were commensal or mutualistic with them. The idea that the cliff flora and fauna followed humans from their native habitats to the new habitats being created by people is central to the Urban Cliff Hypothesis.

Plant and animal species adapted to the cliff environment share a number of qualities that turned out to be crucial for their successful recruitment into the built environment. What cabbages, cockroaches, pigeons, and rats all have in common is developmental opportunism and the capacity to reproduce while very young or very small.[204] Plants such as cabbage, common groundsel, or eastern white cedar can flower and set seed as tiny individuals growing in the smallest of crevices on an exposed cliff face. Yet when these same species are placed into an environment with abundant resources, they have the ability to grow several orders of magnitude faster and larger. The favourable and very rapid response to confinement in a nutrient-rich environment is a characteristic shared by the weeds and pests that plague us in our urban environments today, and the wild ancestors of the crops, horticultural plants, livestock species, and companion animals that we have domesticated or recruited into agriculture.

Of course opportunistic cliff organisms taking advantage of humans and their built environment are only part of the story. The inherent ability of cliff species to quickly exploit improved growing conditions made them particularly suitable for being exploited in turn by the ultimate opportunistic cliff species, humans. In the following sections of this chapter, we will illustrate in more detail how we believe the relationships that we have today with the flora and fauna that surrounds us could have developed. We will begin by briefly discussing the organisms that we do not want in our built surroundings, our urban weeds and pests. Following that we will present several examples of organisms we have welcomed or even actively recruited into our environment.

Weeds and pests

Walled structures large and small are spontaneously colonized by plants and animals native to rock outcrops and cliffs. Prior to its restoration in 1871, the Colosseum in Rome was a naturalized garden of grapes, capers, *Acanthus*, and 418 other herbaceous and woody plant species[205] (more about this in Chapter 6). These species found that the Colosseum walls offered the same growing conditions as the cliffs and rock outcrops where they first evolved. The large sandstone castle at Heidelberg, Germany, is festooned with species of fern and higher plants that normally grow on natural rock faces, and castles and rock walls throughout England are covered in cliff plants.

Besides creating new rock surfaces, caves, and crevices that could be colonized by cliff plants and animals, the construction of buildings began to transform the original landscape in a number of other ways. These included the removal of organisms already present, the alteration of microclimate and natural drainage, the mixing and compaction of soil, and local enrichment in the supply of organic matter. These changes benefited many of the plants and animals listed in Table 1 and offered opportunities to others not generally considered commensal or mutualistic with us. The partitioning of space into units such as buildings, rooms, or bins that segregated food storage from living space provided opportunities for some of these organisms to exploit our unprotected resources. Spaces regularly heated by fire allowed organisms to survive that could tolerate the hostile growing conditions in a rocky substrate[206] but would otherwise be limited by a cold climate.

Built and walled environments, therefore, became attractive to a suite of unwanted organisms including mice, rats, pigeons, cockroaches, and large numbers of other opportunistic species. Most people strongly dislike these commensal organisms. One reason for this is that they feed on our food supplies or waste in darkened rooms, within wall cavities, in ceilings, basements, or storage bins. When disturbed, these animals frequently scurry away in a manner that people find frightening. The same rapid escape from humans is seen in bats and they, too, are feared by many people despite the well-known fact that they are beneficial insectivores. Even if these organisms do not seriously interfere with our success, their presence is an indication that our attempt to control nature has failed.

Human dwellings and dog evolution

Many examples show how the ecological characteristics of human dwellings selected for the development of mutualistic relationships between humans and the species surrounding them. The first and probably the best of these examples emerges from work done on the evolution of the domestic dog.

The close genetic relationship between wolves and dogs is demonstrable by the existence of fertile hybrids between the two.[207] Nevertheless, the transition between wolf and dog in the fossil record appears to be rather abrupt.[208] Wolves occasionally use caves at the base of cliffs for dens, but they clearly exploit a wide variety of other habitats for den selection. We will not make the claim that they are in any way an obligate cliff or cave species. However, wolves would have interacted with humans whenever they attempted to exploit a cave already occupied by humans or the food caches that humans had secured in caves.

Morey (1994) argues that between 20,000 and 14,000 years ago, wolves were acting as scavengers in the same geographical area as hunting and gathering humans. Wolves were therefore competing with humans for the limited supply of energy and space available at the peak of glaciation. Ecological theory predicts that intense competition between two species will result in a divergence of their resource needs, making it possible for the two species to coexist. Coexistence of two species through a differentiation of their resource needs requires less energy than engaging in acts of competition, and is therefore the strategy that is selected for. All alternative strategies involve the mortality of one of the participants.[209] For this reason natural selection in wolves would have strongly favoured the ability to exploit organic waste and debris around human campsites. In other words, dog-like traits would have been selected for within wild wolves.[210] Wolves have always been effective scavengers as well as efficient predators, but the behaviours that facilitate scavenging are very different from those that result in effective predation.

Predation by wolves on other vertebrates, especially ungulates, is much more effective when a pack of animals is involved. In fact, isolated wolves are ineffective at predation on large ungulates. Mitchell et al. (1996) suggest that the wolf–human mutualism may have had its origins in the hunting and killing of horses. Wolves with the first signs of domestication would have been valuable hunting companions to early humans because they would have created a distraction during the attack. The Coppingers (2001), however, point out that modern wolves are timid around other predators and suggest that the wolf–human–horse model is weak. Instead, they argue, the type of wolf–human interaction that was pivotal to dog evolution was the exploitation of organic waste by the wolves around campsites. The mechanism they propose involves natural selection for a short "flight distance" – the distance the canid flees when confronted by a human during feeding. With organic waste from human campsites as an important resource, selection would naturally favour smaller-bodied, smaller-brained, less pack-dependent, and less fearful wolves. Animals with a shorter flight distance would be able to feed more efficiently, which would immediately result in higher reproductive success. In addition to this positive selection pressure, a strongly negative one would apply: any wolves that showed aggressive tendencies toward people would be killed, especially ones with a short flight distance. This combination of positive selection for the ability to exploit waste with negative selection against aggression toward humans would explain the exceptionally rapid evolution of the full array of dog-like traits in wolves.

Supporting evidence for this mechanism comes from a series of elegant experiments conducted by Russian scientists over the past 50 years.[211] These famous experiments involved the development of domesticated foxes from wild ones. The Russian scientist Dmitry K. Belyaev demonstrated that within 40 years, strong and consistent selection against foxes displaying aggressive behaviour toward handlers resulted in the development of neoteny – the persistence of juvenile stages in an adult organism due to incomplete development – in subsequent generations. This neoteny resulted in fox behaviour that was similar to that seen in domestic dogs. Reproductive cycles were also shortened, but more importantly, neoteny brought on a wide array of changes in morphology such as shortened faces, a wider variety of body sizes, and shorter tails, all of them easily identified features that are preserved in the skeletons of the foxes.

In wolves scavenging around early human encampments, equally strong selection pressures against aggressive tendencies would have likely produced similar neotenous traits, resulting in wolves with dog form and dog behaviour (Figure 30). Individual wolves that did not directly threaten people, barked incessantly at the approach of intruders (another neotenous trait), and continued to scavenge organic waste that could be pathological to humans would have been permitted to live in closer association with human settlements because of their ability to carry out important ecological functions: waste management and security around campsites. We have already shown that those campsites were often caves at the bases of cliffs, but the same relationships would have held regardless of where the camp was situated. However, the ecological functions performed by wolves were much more valuable in the cave setting than in any other habitat type. Humans could have simply abandoned dwelling sites in forests or along river banks when organic human waste and waste from hunting activity had accumulated, but this was not an option for caves due to the limited number of such sites. The scavenging and consumption of waste performed by wolves would have lowered the exposure of humans to microbial pathogens and would have made the cave environment more liveable. Hence, humans began to form a mutualistic association with

THE FAR SIDE® BY GARY LARSON

"It's Bob, all right ... but look at those vacuous
eyes, that stupid grin on his face—he's
been domesticated, I tell you."

Figure 30.
Cartoon image from the collection of Gary Larson. Used with permission.

scavenging wolves that were not acting aggressively toward them. If the results of the fox domestication experiments apply to wolves, dogs could have been produced from wolves within a single human generation. Later, when humans began to domesticate other species such as large ungulates, domesticated dogs would have clearly taken on additional value by controlling the movements of livestock and deterring predators. Modern dogs such as the border collie (Figure 31) still display the same behaviours – they are protective of their pack but aggressive to other animals – even though their talents tend to be diverted today to tennis balls, chew toys, and mail carriers.

Figure 31.
The dog (*Canis familiaris*) is viewed by most humans as a fully domesticated mutualistic (and hence benign) animal. Let there be no mistake, the dog is a wolf that has been selected to conduct normal predation and scavenging but stop just short of eating its human host. Photo: D. Larson.

Built structures and the domestication of livestock

Trut (1999) argues that the same array of morphological and developmental alterations described for the wolf–dog transition is also seen in the later domestication of goats, sheep, cattle, pigs, horses, llamas, camels, and guinea pigs, despite the fact that the selection pressures placed on these granivorous or omnivorous animals were probably very different from those placed on carnivorous wolves. The timing of the transition from wild forms to domesticated forms of these species coincides exactly with the timing of the first built human structures. For the period prior to dwelling construction, there is no clear evidence of domesticated animals other than dogs. It is tempting to conclude that for all of these animals, the strong selection for neotenous traits (including juvenile morphology and rapid adult development) occurred at a time when these species were moving from their original habitats into environments manufactured by people.

The ungulate species that were the progenitors of our domesticated livestock inhabited rocky slopes and savanna environments. Their need for access to water courses would have brought them in close proximity to early humans living in caves in river valleys. Evidence shows that at least some of these ungulate species were able to perform ecological functions around human campsites similar to wolves. For example, some ungulates can exploit components of human waste, in particular the salts excreted in human urine. (Note that urine can be exploited only when deposited on a rocky substrate. Urine deposited on soil penetrates completely and is lost to herbivores.) However, at the time of the wolf–dog transition, all other species of plants and animals eaten or otherwise used by humans were still being collected from the wild. Why were goats, sheep, cattle, and horses not recruited into the living space of caves at the same time that wolves were? There are at least two possible explanations. First, some of these species are much larger than wolves. Given the space limitations of cave sites, the addition of tamed cattle or horses would have substantially reduced the space available for people. Second, the defecation behaviour of most ungulates is decidedly different from that of dogs and wolves. The latter rarely defecate in their living space and sometimes bury their feces. Cattle, sheep, horses, and camels, however, defecate wantonly. Living in close association with any of these granivores would have been very challenging as anyone who has ever lived with a cow will know. After humans moved from caves to built structures, most of these constraints on the physical association of people with large ungulates would no longer have applied. Once the ability to construct dwellings for humans had been acquired, structures that allowed for the hunted species to be contained could be contemplated as well. Containment, of course, would have required strong selection against aggressive behaviours in each of the species. Thus the development of neotenous traits in these animals was dependent on the presence of a physical location, a pen, where the animals could be both restrained and protected from predators.

Why were only a selected few of the many available wild species ever domesticated? Diamond (1999) lists a number of reasons why only a few animal species proved suitable candidates for domestication, including the animals' diet, growth rate, mating habits, social organization, and disposition. The origin of the domestic sheep illustrates the variety of factors that interacted to determine whether or not a species ended up being domesticated. The progenitor of the domestic sheep was the mouflon (*Ovis orientalis*), a Eurasian species that lived in rocky talus slope environments. When threatened, mouflons were unable to retreat to higher elevations because these areas were already exploited by three other species with identical geographical distributions. Their exclusive use of lower elevations increased the likelihood of close contact with humans.[212] Mouflons also had multiple offspring and a limited capability to actively defend their young. For these reasons, very young sheep could have been easily captured by humans and either eaten directly or placed into confined settings where they could be raised under greater human control. In contrast, a similar North American species, the bighorn sheep (*Ovis canadensis*), was not domesticated because it differed from the mouflon in several important respects.[213] Bighorn sheep had no direct competitors at higher elevations and thus were able to retreat to less accessible locations when threatened. They also only bear one offspring and defend it vigorously, making it more difficult for humans to capture young animals.[214]

Cats among the pigeons

The evolution of the pigeon, another camp-following species, may have followed a similar model to that of the dog.[215] The pigeon, also called the rock dove (*Columba livia*), originally was a species restricted to rock outcrops. The common image of pigeons surrounding people (Figure 32) merely reflects the opportunistic behaviour of a cliff species: pigeons have adopted the human-built environment because from their perspective it closely resembles their natural environment. Because of this resemblance, all large stone walls or buildings are attractive to them, so much so that the earliest Christian churches were referred to as "houses of doves."

As with wolves, the relationship between the rock dove and early humans was initially commensal. Humans and pigeons coexisted as the pigeon's natural habitats, rock outcrops, cliffs, and caves, were exploited by humans in a variety of ways. Domestication of the rock dove could then have occurred by one of two mechanisms. The first involves the capture of young doves while hunters and gatherers were foraging in rocky terrain. Once returned to the dwellings and raised in the proximity of humans, the squabs most likely remained near the campsites to take advantage of human food caches. Alternatively, rock doves could have discovered human food caches by themselves through their foraging behaviour. Eventually, humans developed a taste for the birds and an appreciation for the agricultural value of pigeon guano, and their relationship with pigeons became mutualistic.

Figure 32.
Flocks of pigeons in Trafalgar Square, London, represent local accumulations of a cliff species that is simply responding to huge surpluses of food that humans offer them. Photo: P. Kelly.

There is evidence that soon after permanent dwellings began to be constructed in Babylon, towers resembling rock pillars were erected in order to attract pigeons.[216] These so-called dovecotes or pigeoncotes, which are particularly common in present-day Iran, Iraq, and Egypt, served the sole purpose of providing habitat for pigeons. The birds were then hunted by people, and the accumulations of guano that built up around the towers formed an important resource to support rudimentary agriculture. The term columbarium in the English language refers both to a pigeoncote and a human burial site. This underlines that the burial of human dead on cliff faces already colonized by doves was a practice established long before buildings were constructed. Rudofsky (1977) asks rhetorically, "Who were the original inhabitants of the rupestral dwellings: doves? men? or both?" He then cites scripture from Jeremiah (48:28) that encourages people to leave the cities and "dwell in the rock like the dove that maketh her nest in the sides of the hole's mouth."

The dove as the symbol of peace and resurrection of the dead has its origins in this ancient connection between cliffs, doves, and human tombs. When doves are released at the opening of the Olympic Games in the modern era, the expression of global peace that they represent reflects the association of *Columba livia* with living and dead *Homo sapiens* on cliffs and rock outcrops.

Like rats and mice, pigeons were initially attracted to cliffs and rock shelters for refuge. Once humans increased the available food supply and constructed new habitat that could be exploited, the populations of these herbivores and granivores exploded. This created huge opportunities for predators of these organisms, among them various species of eagle owl (e.g., *Bubo lacteus*, the African eagle owl) and the wildcat (*Felis silvestris*).

Wildcat skeletons have been found in human-occupied caves dating back 500,000 to 300,000 years.[217] The occurrence of these skeletons in cave sediments does not prove by itself that ancestral wildcat populations were dependent on rock outcrop, cave, and cliff environments. However, it is possible to reconstruct the original environment in which an organism evolved by examining the habitats that currently support the most closely related species. Wild populations of *Felis silvestris* still occur in Scotland, continental Europe, coastal regions of North Africa including Oman and Libya, and western Asia. These populations are frequently restricted to high elevation scrubland, rock outcrop, and escarpment settings where the pressures from human disturbance are low.[218] This retreat of organisms to ancient ancestral habitats when they are ecologically stressed fits in entirely with the idea of cliffs as refuges.[219]

For hundreds of thousands of years, then, wildcats had likely been co-occurring with pigeons, rodents, and early humans in places such as the abundant limestone outcrops along the Nile River and on the perimeter of the Mediterranean Sea. The associations that formed between these organisms may therefore be very old.

When early caching of food in caves drastically increased the opportunities for rodents and pigeons, the human inhabitants would have welcomed the presence of wildcats that reduced the densities of granivore populations. Besides destroying food caches, rodents also bore parasites and carried bacterial diseases that were potentially threatening to human populations crowded into caves and rock shelters. Hence, cohabitation in space or time with wild or domestic cats would not have been selected against, especially considering other characteristics of wildcats that made them suitable for sharing human cave dwellings: wildcats are too small to be a threat to humans, and like wolves, are inclined to defecate outside their den.

Once the first built structures appeared and outbuildings were constructed specifically for food storage, such as the ancient granary at the base of the Nano Escarpment in Togo,[220] shown in Figure 33, the rodents and birds that took advantage of the enormous concentrations of grain and other foodstuffs would have presented a serious problem had it not been for the wildcats now fully accustomed to hunting within an environment occupied by humans. The size of the cat populations, as well as their contact with humans, undoubtedly increased by many orders of magnitude at this time, about 5000 years ago. At what point were cats domesticated? That is difficult to answer. Unlike the transition from wolves to dogs, that from wildcats to domestic cats is not accompanied by any morphological changes that can be seen in the fossil record. However, at Kalamakia Cave, Greece, wildcat skeletons are seen in the fossil record throughout the rapid transition from wolves and ibex in the middle Pleistocene to dogs and goats in the Holocene.[221] The domestication of cats, then, may have taken place simultaneously with these other species.

It can be argued that since cats were the most efficient predators on the rats, mice, and pigeons that competed with humans for food stored in the earliest built structures, civilization could not have developed as well or as rapidly in their absence. The early Egyptians clearly had a sense of the central importance of cats to their economy and culture: the elevation of cats to the status of gods in ancient Egypt is well known and often attributed to the central role that they played in reducing the risks to the stored grain crops. On the other hand, their survival (unlike that of scavenging dogs) was not directly dependent on food being provided by humans, since they remained able to exploit both natural and human-modified environments. Even today cats are capable of an exceptionally rapid transition to feral form and therefore do not really need the protection of humans as long as rodent and bird populations are large. This asymmetry in the cat–human mutualism has persisted to the present day, as every cat owner knows: cats appear to be less dependent on us than we are on them.

Figure 33.
Mid 18th-century grain storage bins constructed at the base of the Nano Escarpment in Togo. Foundations of similar structures exist at the site dating back to the Paleolithic (Posnansky 1980).
Photo courtesy of Dr. P.F.L. deBarros. Used with permission.

A repeated origin story

In the previous sections we have shown how the domestication of dogs, livestock, and cats might have arisen from the close association of humans with other organisms that used the same environment: cliffs and talus slopes adjacent to river courses and surrounded by savanna. We believe that the same general principles apply to most of the plants and animals that surround us today (see Table 1). Central to the Urban Cliff Hypothesis is the view that the same series of selection pressures that led from wolf to dog applied to the entire flora and fauna.

We can now imagine one possible sequence of events that led to the simultaneous development of built structures and the urban and agricultural biota. According to Vavilov,[222] there was an intermediate step in the transition from cave dwellers that hunted and gathered in areas surrounding the rock outcrops to humans that practised agriculture and lived in built dwellings far away from the cliffs: the growing of plants in sloping areas close to the caves where waste had accumulated and water ran naturally. These places would have been attractive to animals. The next step, according to one of the possible scenarios, was the building of retaining walls on the slopes. With the help of these walls, animals could be confined and protected, and backfilling of the slopes behind the walls created terraces where plants could be stored or grown. The construction of terraces on slopes at modest elevations is thus viewed as an early step in agriculture, not a late one imposed on people by competition with farmers on the valley floors. In an early *National Geographic* article, Cook (1916) describes at great length how early Peruvian terrace agriculture exploited sloping terrain and small streams at high altitudes to create some of the most efficient agricultural systems ever known. Potatoes, corn, beans, squash, and manioc (also known as cassava, a tropical New World plant with a large starchy root) were all grown in these terrace systems. Each individual terrace consisted of a rock wall constructed on a talus slope that was then meticulously backfilled with small stones, rich organic debris, and manure. Such retained soil represents nothing more than a highly refined waste site as described by Vavilov.

Pigott (1965) argues that wild plants and animals may have followed humans from one site to another as their offspring were collected and either eaten or confined. This is exactly the same mechanism that Coppinger and Coppinger (2001) have proposed for dogs. Many of the species recruited into agriculture could easily have started off as commensals or contaminants in crops that were actively harvested or cultivated.[223] Budiansky (1992) points out that many of the plants and animals brought into cultivation were on the verge of postglacial extinction from overhunting at the time they were recruited into agriculture. For most of these organisms, population growth exploded as they were brought into environments built by humans. The same idea emerges in Eisenberg's (1998) book *The Ecology of Eden*. Eisenberg also argues that humans attract opportunistic species because we are an opportunistic species ourselves.

Some have contended[224] that the repopulation of Europe following deglaciation was carried out largely by particularly successful groups of

Neolithic farmers who both replaced and were absorbed by existing cultures. Although written documents do not appear for another 5000 years, the spread of early agriculture appears to have depended on people effectively sharing their cultural tools and architecture while also retaining the ability to exploit naturally occurring cave and cliff environments.

A short section of text from Coppinger and Coppinger (2001) is worth quoting to cap off this section of the argument:

> Late in the Mesolithic period (fifteen thousand years ago), hunter-gatherer people built stone villages along savannah-like game trails in what is now Israel. These people were called Natufians. They built the first permanent settlements (that we know of, so far) near reasonably constant sources of food. In Namibia, I once saw an example of what these places must have been like. There was an outcrop of rocks in the middle of a grassland. In the rocks was a little gorge with a waterfall and pool. The Namibians hid under the overhangs, waiting for the grass-eating game to come to the only water in the area. Since the game had to come there, they could establish a permanent ambush. While idling away the time, they drew graffiti on the walls. Saving energy by letting the food come to you rather than chasing after it is a good survival trick. It's like sitting in a dump waiting for food to arrive. At my bird feeder is a feral cat that does the same thing.

The chapters so far have shown that cliffs, caves, and rock shelters provided the optimal habitat for early humans and an array of other species throughout most of the Pleistocene. About 40,000 years ago humans evolved the capacity to construct shelters. In their attempts to retain the useful features of caves while eliminating the drawbacks, they inadvertently created new habitat for the opportunistic cliff species that had shared their original environment. Some of these organisms still take advantage of us today in the form of our weeds and pests, while we have recruited others into mutualistic relationships as our pets and agricultural species. This leads us to the topic of architecture. As we will see in the next chapter, its history involves the continuous refinement and expansion of the themes that were so successful in natural cliff environments and early built structures.

164   Rudofsky (1977).
165   Goodall (1988), Diamond (1992).
166   McGrew et al. (2003).
167   Pfeiffer (1972).
168   O'Sullivan (1994).
169   Smith (1992), Pfeiffer (1972).
170   Smith (1992).
171   Nesse and Williams (1996).
172   In some rare settings, such as the Amud Cave in Israel (Johanson and Edgar 1996), a perennial stream exists in the area. As a consequence, this site was used extensively between 50,000 and 40,000 years ago.
173   Courbon et al. (1989).
174   Guidoni (1978).
175   Meyer (1969).
176   Binford (1984).
177   Bordes (1972).
178   Otte et al. (1998).
179   Johnson (1911).
180   Waddell et al. (1994), O'Connell and Korff (1991).
181   Baring-Gould (1911), Kempe (1988).
182   Carr (1977).
183   Bordes (1972), Klein (1973), Johanson and Edgar (1996).
184   Gladkihm et al. (1984).
185   Thorpe (1996).
186   Hoebel (1966).
187   Carswell et al. (1981)
188   Trachenberg and Hyman (1986).
189   Nuttgens (1997).
190   Rudofsky (1977).
191   Rudofsky (1977).
192   Nuttgens (1997), Samuels and Prasad (1994).
193   Jacobs (1968), James (1965).
194   Tilley (1994).
195   Williams (1930).
196   Carswell et al. (1981), Perkins (1909).
197   Moore (1960).
198   Toy (1985).
199   Samuels and Prasad (1994).
200   Crouch (1985), Nuttgens (1997), Trachenberg and Hyman (1986), Rudofsky (1977).
201   Cook (1994).
202   Darlington and Dixon (2000), Darlington et al. (2001).
203   Carney (1995).
204   Larson et al. (2000a).
205   Quennell (1971).
206   Spirn (1984).
207   Morey (1994), Vilà and Wayne (1999), Coppinger and Coppinger (2001).
208   Smith (1992).
209   Keddy (1989).
210   Coppinger and Coppinger (2001).
211   Trut (1999).

212 These three species were ibex (*Capra ibex*), chamois (*Rupicapra rupicapra*), and mountain goat (*Capra hircus*).
213 Carr (1977).
214 Bassett (1996).
215 Johnston and Janiga (1995).
216 Rudofsky (1977).
217 Gamble (1999a).
218 Easterbee et al. (1991), Hubbard et al. (1992), Kitchener and Easterbee (1992), Lozano et al. (2003).
219 Larson et al. (1999).
220 Posnansky (1980).
221 De Lumley and Darlas (1994).
222 Löve (1992), translation of Vavilov (1926).
223 Vavilov (1926) as translated by Löve (1992), Sauer (1993).
224 Bogucki (1996), Semilo et al. (2000).

CHAPTER FIVE

Home, Sweet Home:
The Human Attraction
to Cliffs

There is little argument that humans are deeply and instinctively attracted to cliffs and rock outcrops. Give people a perfectly flat section of terrain, and they will hunt for the first elevated rocky point of land they can find. This instinct may be no different from that found in insects, which usually climb to the highest point on a stick before they take off, but there may be more to it than that. In this chapter, we will begin by showing that this fundamental attraction has existed throughout our history right up to modern times, and that cliffs, rocks, monoliths, or mountains – and later on, built structures made to resemble the most spectacular forms of these – have been considered spiritually charged or holy places in many faith systems all over the world. We will go from there to explore the reasons *why* rock outcrops, cliffs, and large built structures create an environment that is intensely spiritual. We will invoke E.O.Wilson's Biophilia Hypothesis to explain our attraction to these structures and will show that we are attracted to rock because it has been our home since our earliest beginnings.

Rock in history and architecture

Rock outcrops, cliffs, caves, and mountains were centrally involved in many religions and cultural traditions throughout prehistory, an attraction that we believe persists to this day. The idea that rock outcrop sites have had spiritual significance since the time of early humans is supported by a great deal of early archeological evidence.[225] Modern writers, too, report that great reverence for cliffs and rock outcrops still exists in many cultures. For example, in Mali, in sub-Saharan Africa, the Dogon peoples honour rock outcrops along the Bandiagara Escarpment as dwelling sites enriched with protective and spiritual functions.[226] Similarly, the Mayans of Belize considered caves in natural escarpments and the naturally produced crystals within them to be places of extreme religious significance: to them these sites were not just physical places but rather living manifestations of spiritual power, including entrances to the underworld, sources of drinking water, sources of virgin water, sites of human burials and cremations, locations with art galleries, and places of refuge.[227]

References to caves as repositories of animal spirits or to rocks as dwelling places of deities are common in many mythologies, and even a modern volume on the greatest caves of the world uses religious terms to describe the largest natural cave galleries.[228] Bernbaum (1997) speculates that the sacredness of the Himalaya Mountains to Hindus is reflected in the meaning of the word *Himalaya* itself. It is translated in ancient Sanskrit as "abode, or home of snow" of a god who placed himself in a "paradise sparkling with streams and forests set beneath beautiful peaks." The Collauas peoples of Peru, named after the Holy Mountain (Collaguata) from which they emigrated, still revere their ancestral cliffs today, to the point that they continue to transport their dead back to the holy mountain to bury them in caves.[229] A portion of the population of Easter Island – an island famous for its immense statues of large faces carved out of rock[230] – once lived in caves and was part of a religious cult that revered cliff-dwelling birds.[231] In many cultures the rocks themselves are thought to be alive. The Buryats of central Mongolia, for example, believe that their most important shamans are transformed into cliffs after their death.[232]

The use of escarpments and later of their large built equivalents for religious ceremonies across all continents and times also suggests that these places are rich with meaning.[233] Rock art, pilgrimages to rock outcrops, and initiation rites involving cliffs are exceptionally common in all pre-industrial cultures, and the built escarpments known as cathedrals and city halls around the world are still used for special occasions and ceremonies. In the epic Poem of Gilgamesh,[234] the rocky slopes of the mountains sustain Enkeidu before his relationship with Gilgamesh develops, and the rock outcrops, precipices, and cedar forests guarded by Humbaba provide the setting for one of the central conflicts of the poem.

We turn now to look in more detail at several manifestations of our attraction to rock: rock art, cliffs used in sacrifice, and cliffs used for interment of the dead. This will lead us to examine how religious images and later buildings with religious purpose were carved into rock, and how this spiritual significance was transferred from cliffs to built structures made to resemble cliffs.

Rock art

Many people have been struck by the consistency of the appearance of art on cave walls and the almost universal suggestion that somehow the rock outcrops and cliffs above the caves have spiritual or educational value. The vast literature on this topic[235] confirms that human cultures worldwide have valued the idea of recording artistic and historic images on the walls of caves and rock shelters. Rock art in the form of carefully executed paintings or carvings on cave walls has occurred at sites such as Niaux, Chauvet, Cosquer, and Lascaux in France for more than 35,000 years. It is found on vertical rock walls throughout many different parts of the world, including southern and Saharan Africa, Australasia, Europe including western Russia, and the Americas.[236] Many believe that the function of the art was to educate hunters about the behaviours of game animals; others argue that the appearance of human-animal chimeras suggests a deeper spiritual meaning.[237] Lewis-Williams argues that Paleolithic rock art often served as a membrane that connected people to the world of spirits that flowed from the real world through rock shelter sites to the underworld and back again.

We can gain valuable insight into the possible motivation of the prehistoric cave artists from interviews with present-day rock artists in North America (Grand Canyon), South Africa, and Mesoamerica,[238] all of whom attest to the spiritual significance of the artistic objects added to the caves. The cave and rock art drawn by living persons today is sufficiently similar to that drawn in the past for us to infer the intent of those long dead Meso- and Neolithic artists: to be able to physically or mentally return to sites that offer both life and death and to prepare permanent images to honour these forces of nature.

The existence of artistic artifacts (paintings and carvings) on cliff and cave walls that started about 35,000 years ago[239] blends imperceptibly into the cliff and wall paintings of Bhutan or the ancient and more modern frescoes of Greek and Italian courtyards. Handprints painted with the spit-paint technique have been found in natural cave settings older than 30,000 years as well as on the walls of the 8000-year-old city of Çatal Hüyük in Turkey.[240] This continuity over time suggests that the cave paintings at Lascaux were functionally identical to paintings in the Sistine Chapel or modern art galleries.[241]

Not only were cliffs and their built analogs often decorated with art, but from the Renaissance onward visual art has with striking frequency featured depictions of these landscape structures. Rock outcrops, escarpments, caves, and mountains were all central themes in the development of landscape art in Italian Renaissance painting. Turner[242] reviews the history of the landscape art of this period and presents images of more than 130 works (such as the Mona Lisa) with themes that express attitudes to landscape and the spiritual world. Cliffs and other types of extreme topography figure prominently in about one-third of these works, a far greater number than one would expect given the actual frequency of such scenes in the area.

Not only paintings but also other forms of artistic expression are often linked to cliffs and rock.[243] The evidence is compelling in some cases that the rock carvings in caves, especially in caves with stalactites, may have resulted in the discovery by Neolithic peoples of the beauty of musical tones by using the rocks as lithophones.[244] Lewis-Williams presents the familiar argument that cave art allowed Paleolithic peoples to make connections with the world of spirits, but he also notes that music has exactly the same effect. He adds that

> the combined effect of the underground spaces, altered states of consciousness, mysterious sounds, the interplay of light and darkness, and progressively revealed, flickering panels of images is, in one sense, easy to imagine, but, in another sense, the impact of such multi-sensory experiences on Upper Paleolithic people probably exceeded anything that we can comprehend today.

In some limestone regions of Vietnam, ancient stalactite lithophones have been found in human-occupied caves that also had purpose-built seven-tone gamelans.[245] In arguing that these lithophones functioned exactly like the gamelans, Hood (1980) seems to assume that gamelans were constructed prior to the use of stalactite lithophones. We suggest that the sequence of development could easily have been the reverse: that Paleolithic peoples "discovered" the capacity of their rock shelters to produce and shape musical tones and that this discovery, once learned, was passed down from found objects to the built environment. Unfortunately, exact dates of construction or use of lithophones in rock shelters have not yet been determined. Dams (1984, 1985), however, reports thick accumulations of calcite on top of the black and red ochre markings on the striking surfaces of lithophones, suggesting the passage of great periods of time.

If we rush from the distant past to the historical period, we see a continuation of the same trend. For example, the construction of theatres in rocky terrain has spanned the period from ancient Greece (theatre of Apollo) to modern-day Colorado (Red Rocks Amphitheater). In these settings, the combination of refuge seating and prospect view is provided by the rocky canyons within which the theatres are situated. None of this exploitation of rock for artistic or spiritual purposes would be expected if people simply used rock outcrops, cliffs, or caves according to their commonness in the landscape.

Sacrifice and scapegoats

The idea of sacrificing the life of an animal or another human is extremely powerful and widespread in humanity. Civilizations in places as diverse as Africa, Asia, Europe, and Central America are believed to have practised human sacrifice at the summits of natural rock outcrops or mountains and later on platforms, monuments, or pyramids built to mimic the natural shapes of such familiar landforms.[246] Human sacrifice and interment in caves dates possibly to the Neanderthal period, although recent thinking restricts the ceremonial interment of the dead to modern humans.[247] The interment of living children in the foundations of buildings is reported for Druids in Britain and many other countries in Europe, the Mediterranean, and India, and has continued in some cases through to the modern era. These burials presumably allowed the souls of the children to permanently protect the sites from evil spirits.

Cliffs are centrally involved in the original legend of the scapegoat.[248] This legend relates a mechanism for small social groups to resolve conflicts imposed partly by external circumstances and partly by their own normal selfish behaviour. When these groups felt the need to blame one another for their unhappiness, a goat would be selected to represent the embodiment of the psychological problems that the group was experiencing. The now demonized goat was then led away to a high escarpment and driven off the cliff edge to its death. The death of the scapegoat as a demon was believed to be the death of the problems. The portrayal of the Devil or Satan as a cloven-hoofed creature with short horns and linear irises appears to be a manifestation of these demons as goat-human chimeras. In fact, the Hebrew word *Azazel* refers to both the name for a scapegoat and the devil. Girard (1989) argues that the death of the scapegoat allows for continued beneficial social development, and that not just Christianity but all of human development is based on variations on the theme of scapegoat persecution.

Interment of the dead

Burial of the dead in caves at the bases of cliffs is a very ancient practice that appears to be restricted to modern humans. The behaviour was carried out by both *Homo neandertalensis* and *Homo sapiens*,[249] but there is a growing consensus that Neanderthal burials were probably more practical than spiritual.[250] When space was insufficient for burial in caves at the bases of cliffs, natural solution pockets or crevice caves were enlarged to create burial chambers in the rock. Such activities took place all through the Fertile Crescent and have continued to modern times in the practice known as "hanging coffins" on cliffs in Southeast Asia.[251] Caves carved into cliff faces and then used to secure human remains have also been reported from Carlsbad, New Mexico,[252] and caves in low-elevation escarpments in Peru were used for centuries to house the dead of revered persons.[253]

To this day, the most celebrated members of our communities are interred at the bases of large memorial buildings whose main design objective is to bring recognition to the life of the person.[254] In Britain, similar status is bestowed on people by interment in Westminster Abbey, which can be considered an outstanding example of a built escarpment riddled with hundreds of below- and above-ground permanent chambers.

Graven images

Holy or spiritually charged images or structures carved directly into rock outcrops and cliffs are extremely common around the world and across the cultures of Buddhism, Islam, Hinduism, and Christianity. Such objects of religious significance range from inscriptions and carvings to statues chiselled out of the rock to shrines, tombs, monasteries, and churches built into and integrated with the natural escarpments. On Mount Kailas in Nepal, various forms of art and symbolic language are painted or carved into rock faces as a form of homage to this mountain, which is revered as the Mountain of God.[255] In 516 B.C., the Persian king Darius the Great selected a site at Behistan, present-day Iraq, to record 13 panels of text carved into a towering cliff face known as the "Mountain of the Gods." These carvings played a central role in deciphering cuneiform script. Similar inscriptions have also been found in Syria.[255] Prior to their destruction by the Taliban-controlled government of Afghanistan in 2001,

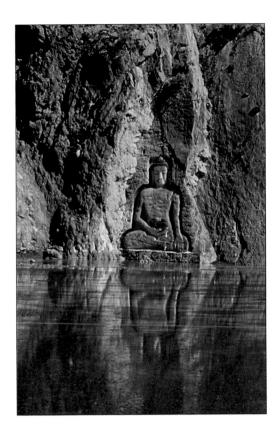

Figure 34.
Rock-cut temples and statues of the Buddha are found throughout China, including Tibet (as shown here), Afghanistan, and other Asian countries. Such statues rise in some cases to more than 50 m in height and have dwelling sites for monks behind the statues. Photo: W. Witt.

towering images of the Buddha, which had been carved into sandstone cliffs, looked over the town of Bamiyan, near Kabul. Such statues, found throughout China and other Asian countries (Figures 34, 35), are thought to have been carved by Buddhists from 1500 to 2000 years ago in locations that would generate reverence for the philosophy of the Buddha.[257] Monks' dwellings were carved into the same cliffs, and some of them are still occupied (Figure 36). In the Valley of the Kings, Egypt, and at Petra, Jordan (Figures 37, 38), large natural rock outcrops and cliffs were converted into gigantic symbols of the faith systems of different peoples.[258] Shrines or monasteries carved directly into escarpment faces are found through all of the limestone regions of southern China[259] and elsewhere in Asia as well as throughout Europe.[260] The Ajanta Caves in India, Valley of the Kings in Egypt (Figure 37), the Dome of the Rock in Jerusalem, Meteora in Greece,[261] le Citadeole des Baux in France, and a small Christian church called Felsenkirche that is built directly into a cliff face in Idar-Oberstein, Germany, are just a few examples that demonstrate this integration of a built structure with a naturally occurring cliff face.

Figure 35.
Rock outcrops and cliffs in Bhutan are places of great spiritual significance.
Here, a painting of the Buddha is engraved into a rock face.
Photo: L. Melville.

Figure 36.
These cliff temples and
residences for monks in
Bhutan reflect the spiritual
importance of rock to them.
Photo: L. Melville.

Figure 37.
Natural rock escarpments, carved tombs, dwellings, steps, and rock-block
construction are juxtaposed in the Valley of the Kings, Egypt.
Photo: J. Lundholm.

Figure 38.
At Petra, Jordan, natural rock
escarpments were modified to
form buildings of great
spiritual significance.
Photo: J. Lundholm.

Figure 39.
Tilley (1994) argues that stone megaliths throughout Western Europe
represented methods of establishing the presence of special "places" in
landscapes. It is argued that such stone constructions were recapitulations of
stone monuments made by humans since 5000 to 8000 years before present.
Photo: D. Larson.

Altars of stone

Like rock outcrops and mountains, free-standing structures constructed of rock (such as monoliths, pillars, and altars) appear to be centrally involved in many religions and cultures. Even though pillars of wood have surrounded many Druid sites in Britain, the pillars of stone seem to have been the central focus points at sites such as Stonehenge (Figure 39). The book of Genesis (28:18-22) in the Jewish and Christian scriptures refers to Jacob as having been ordered by God to erect a permanent stone pillar to represent the house of God. The construction of stone pillars, monoliths, obelisks, and tablets that are holy or spiritually charged has continued throughout recorded history,[262] and such structures are invariably interpreted as sites of great spiritual contact and importance.[263] The most sacred object in all of Islam is a dark-coloured stone in the outer wall of the Kaaba (house of God) in Mecca,[264] believed to be the only remaining part of the temple erected by Abraham. Collections of prayer stones placed against the base of rock cliffs are considered sacred objects in Bhutan (Figure 40).

Buildings with spiritual significance have been constructed for millennia to duplicate the grandest of naturally occurring escarpment environments. Temples, churches, cathedrals, and tombs around the world are designed to engender the feelings of protection, power, and permanence that are reminiscent of large caves.[265] The Amernath Cave, located at the base of a cliff in Kashmir at 4000 m altitude, is regarded in Hindu culture as the first dwelling of the god Siva. In an early *National Geographic* article, Jessop (1921) describes an expedition to this cave:

> We had not realized what a visit to this cave means to a Hindu. The Christian has his Jerusalem, but he goes there as to a historic city . . . But to the Hindu pilgrim, Amernath was the first abode of his god Siva, the destroyer who ushers in another life, and the word means life that never ends; so that the devotee who enters the inclosure treads upon holy ground and receives the gift of everlasting life.

The temples carved out of solid cliffs at Ellora and Ajanta in India were intended to be human-built copies of the natural cave that gave birth to Siva. Wentzel (1953) states:

Most Ellora temples belong to Siva, for he was the favorite among the Brahmans who built them. Siva's colorful and complex personality delighted intellectuals of the Hindu priesthood. As the destroyer, he removed beings who became weak and useless. By refusing to destroy, he became the protector of the strong and the useful. Siva was the divine dancer who could interpret the mathematical law of the universe in 108 different movements, and he was the Lord of Knowledge, the centre about whom the universe revolved.

The connection of cliffs and caves to a life-giving and life-taking force is by no means unique to Hindu culture. For example, Stuart (1981) has suggested a similar interpretation of Mayan cave paintings and sculpture. Biologists reading the above passage might conclude that Siva symbolizes natural selection and that the birthplace of Siva, a cave in a cliff face, therefore represents the location where natural selection originated. We can then argue that for Hindus, the construction of rock-cut temples at Ellora and elsewhere duplicated the setting in which natural selection was given birth.

Figure 40.
Prayer stones are commonly inserted into crevice caves
at the base of cliffs and rock outcrops in Bhutan. Photo: L. Melville.

The pyramids built in Egypt and Central America represent prominent meeting places or sites that permanently honour the dead, as well as places enriched in spiritual power. James (1965) comments:

> When the Sumerians settled on the alluvial plains and marshes of the Euphrates valley, they were doubtless compelled to raise their buildings above the water level on artificial foundations secure against the periodic floods. But apart from this practical necessity, being in origin a mountain people they were accustomed to worship their gods in sanctuaries elevated on hills. Therefore, they continued this tradition placing their temples on mounds and platforms, the Ziggurat being given a lofty elevation symbolizing the mythical mountain of the world, representing the structure of the universe.

Like the massive carved sanctuaries of the Buddhists at Ellora and Ajanta, cathedrals of the Christian era represent cavernous buildings that house pigeons, the human dead, and holy spirits. All of these places help promote and solidify social harmony and direction. Morrish (1996) claims that all of the best modern urban architecture directly borrows themes from mountains, steep-sided mesas, and other rock structures in nature (Figures 41, 42). He argues that such natural "places" are divine resources that we have lost awareness of. Greater respect and connection can be made by constructing buildings organized around these themes.

So far in this chapter, we have shown that the attraction humans feel for rock outcrops, cliffs, caves, and their built substitutes is ancient and universal, as is the spiritual feeling they engender and their connection with a life-giving and life-taking higher being. In the remainder of the chapter, we will ask the question why this is so.

Biophilia of rock

Why do rock outcrops and cliffs sustain so much more attention, attraction, and feeling from humans than one would expect based on their immediate utility or their frequency in the landscape? As we have shown in Chapter 2, rivers, riverbanks, forests, and open plains are all much more abundant than rock outcrops. Human faith systems seem to involve rocks, cliffs, mountains, or their built substitutes far more frequently than could be explained by chance alone. Great spiritual significance has certainly been

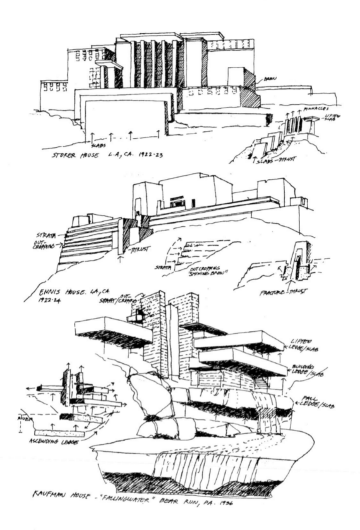

Drawing 35.
THE SHINING BROW

Figure 41.
The connectedness of the human-built environment and natural rock outcrops, mountains, and escarpments. Vertical and horizontal rock outcrop elements were incorporated by Frank Lloyd Wright into some of his most famous buildings including Falling Water, Bear Run, Pennsylvania, and Storer House, Los Angeles, California. Illustration taken from Morrish (1996). Used with permission.

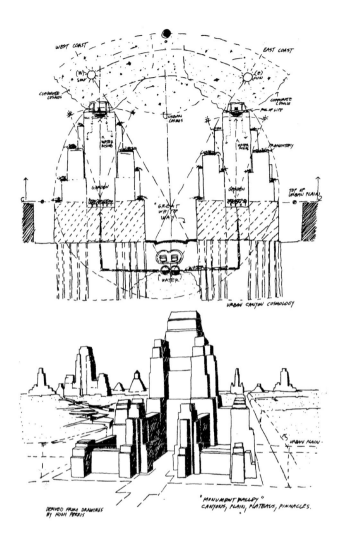

Drawing 29
MONUMENT VALLEY

Figure 42.
The connectedness of the human-built environment and natural rock outcrops, mountains, and escarpments. The drawing emphasizes the commonality of natural themes in Monument Valley, Utah, with urban themes in modern cities.
Illustration taken from Morrish (1996). Used with permission.

attached to some rivers such as the Ganges in India, but the same is not true for all rivers, and even the Ganges itself is venerated as the water that flows from the sacred mountains. Since rivers provide an endless supply of water for drinking and irrigation, one would expect the greatest spiritual value to be placed on the rivers themselves rather than on the place where the water originates. Equally, why not revere open plains richly abundant with grain and animals, or majestic forests that supply fuel and building materials and contain trees that, from a human perspective, live forever?

We cannot know the answer to this question. But as we have shown in the first part of this chapter, the pattern is so universal that we believe it to be primitive and fundamental. Consider the enormity of the effort to hand carve the rock temples of Ellora or Petra or the fantastic temples of Bali;[266] there can be no doubt that the engineers and planners had an exceptionally strong incentive to do the work. We will argue below that the pattern itself has not disappeared but modern society has simply dressed it up in different coverings.

The attraction to rock has been recognized in the literature, but there have been few attempts to provide an explanation for it. *A Pattern Language*,[267] for example, points out that rocks, stones, and mountains become holy places for humans but offers no reasons why this might be so. Elsewhere it has been suggested that individual rock outcrops or large stones have significance within landscapes because they represent discrete "places" that would otherwise be difficult to define spatially.[268] The visibility of such landscape elements is said to generate feelings of control and certainty, and these feelings are enhanced by being able to name a site.[269] However, one could easily counterargue that one can name streams, valleys, or grasslands, too.

We are proposing that the attraction we feel for cliffs is based on the ecological function that cliffs performed for us in prehistory. As we have shown in Chapter 3, rock outcrops, caves, cliffs, and mountainous areas provided us with optimal habitat and refuge and were a source of all of the resources we required. In the 1920s, Vavilov (1926) wrote:

> Primitive man feared, and still fights up to this day, the wet Tropics with its unmanageable vegetation and tropical diseases, in spite of the fact that the wet Tropics with their fertile soils occupy one third of the world's landmass. Man settled, and is still settling down along the edges of the tropical forests, but the mountain areas of the Tropics and Subtropics offered more

favourable conditions for the original inhabitants in the sense of a warm climate and an abundance of food as well as the possibility of life without clothing.

The universal human attraction to cliff and mountain settings could easily be a result of the early nurturing that these habitats provided our ancestors. Certainly Cameron (1950) came to this conclusion about the particular site at Behistan, Iran, chosen by Darius the Great for his massive autobiographic rock inscription. Our species has been associated with these environments for a long time, environments that have always represented the safety, security, and resources of "home." We propose that over time, this feeling has been shaped and reinforced by natural selection. This idea is supported by invoking suggestions from evolutionary psychology that fall under the topic of biophilia. We turn now to a discussion of the Biophilia Hypothesis as put forward by E.O. Wilson (1984), after which we will explore the connections between his hypothesis and ours.

Most agree that our attitudes, beliefs, and feelings are taught to us by our family and friends and reinforced by societal institutions and communicative media. In recent years, however, an alternative idea has been presented: that attitudes and feelings can be the product of evolution. Wilson was the first to popularize these ideas under the heading of biophilia.

According to the Biophilia Hypothesis, the attraction that humans feel for the natural environment has a biological basis and is driven by the same evolutionary processes that have shaped our anatomy and physiology.[270] A wide variety of environmental and psychological studies support the claim that humans have a deep instinctive preference for certain aspects of their environment.[271] Most subjects of such studies favour scenes of savanna that include sources of water over other habitats such as forest, tundra, and desert.[272] Reacting to scenes of savanna, male subjects display a preference for prospect views that allow a wide vista to be seen. In contrast, female subjects display a greater interest in images showing refuges that offer protection. Even more interesting than these results, from our point of view, are the procedures commonly used in experimental studies of human habitat preference. When subjects are asked to examine photographs of different habitat types, the images are often pre-selected specifically to exclude the most picturesque images of mountains (Figure 43), cliffs, or waterfalls (Figure 44).[273] Excluding scenes of spectacular beauty or extreme topography seems natural to the investigators who construct the experiments. They feel that such images would bias the results of their investigations, since people are more likely to be drawn to them! From our point of view, however, such a bias merely confirms the point we are trying to make. Referring to the reverence for mountains in Buddhism, Govinda (1970) writes:

> Nobody has conferred the title of sacredness on such a mountain, and yet everybody recognizes it; nobody has to defend its claim because nobody doubts it; nobody has to organize its worship, because people are overwhelmed by the mere presence of such a mountain and cannot express their feelings other than by worship.

Is there empirical evidence that present-day humans exhibit a stronger interest in rock outcrops, cliffs, and caves than one would expect by chance? We asked this question in our book *Cliff Ecology*.[274] There we showed that cliffs are used far more extensively in commercial advertising than one would expect if there was no human bias in favour of images of one habitat type over another. Two hundred and twenty items of commercial advertising that included photographs of natural landscapes were located in four major publications over two years. The results showed that 31% of the images had cliffs in the foreground or in the background, while only 4% showed forest, 19% lake or ocean, 12% open fields, and 34% all other habitats combined. Given the evidence presented in Chapter 2 that rocky habitats represent vastly less than 1% of the vegetated surface area on earth, one would expect to see images of cliffs only rarely in graphic images. A similar argument can be made for the huge number of cliffs, rock outcrops, and escarpments that form the focal points of parks and natural areas around the world. If cliffs held no greater attraction for us than other habitats, then the abundance of cliffs that are protected in the form of parks or nature reserves should only reflect the relative abundance of these features in the landscape. Considering the relative rarity of cliffs and vertical rock outcrops in nature, there can be no question that there is far more public interest in these places than one would expect if no preference for these locations was present.

The great human attraction to cliffs is not inconsistent with the conclusions of others that humans are attracted preferentially to savanna habitats.[275] This is because most African savannas are eroding plains dissected by river courses that carve into the underlying rock, thus creating cliffs. As we discussed in Chapter 3, the foraging opportunities offered by

Figure 43.
Spectacular escarpment scenery in the Usambara Mountains, eastern
Tanzania. Such scenery is present in many countries in east Africa.
Sites such as these are routinely excluded from studies of landscape
preference by humans, on the basis that most people would
obviously prefer them over others. Photo: J. Lundholm.

savannas are greater by many orders of magnitude than those offered by
other types of environment – but they are offered not only to humans but
also to other granivores and carnivores. Hence, perfectly open and flat
African savanna is hostile to organisms whose survival is dependent on

Figure 44.
Spectacular waterfall at Ein Gedi Nature Reserve. This site is close to
Qumran, where the Dead Sea Scrolls were discovered. The similarity of this
site to Figure 18 is remarkable. Photo: J. Lundholm.

finding refuge. Once points of refuge are present within a savanna, however, all of its disadvantages disappear. The evolutionary psychologists whose experiments have established the pattern of human preference for savanna rarely attempt to identify the particular features that provide refuge within savanna habitat.

Among the few who have are Heerwagen and Orians (1993). They show that before humans acquired the ability to control fire, the refuge value of savanna was enhanced by the presence of cliffs with caves at their base. They also point out that predators were common on open grassland and that rock outcrops offered prospect views and refuge sites *at the same time*. We have already discussed this in Chapter 3, and have shown that cliffs and the caves within them have sheltered and nurtured at least three species of humans for several hundred thousand years, an enormous amount of time in the context of the historical record of humanity.

In addition to representing a source of predictable safety in people's lives, cliffs have been used on many continents and throughout evolutionary history as a lethal tool to harvest large game by driving individual animals or whole herds of them over the edges of escarpments.[276] For perhaps 250,000 years, humans would have associated their own survival with the death of other organisms killed by falling from the top of a cliff or rock outcrop. As intelligence and an awareness of "self" evolved, so would the awareness of our own mortality.[277] This awareness of mortality may have been maximized in rock outcrop settings where life and death are finely balanced. This is exactly the essence of the Hindu god Siva, the destroyer who is revered for his ability to understand the function of the universe and dispense control over life and death.[278] Bourassa (1991) argues that rock outcrops and cliffs are considered sublime landscapes precisely because they conjure up images of immediate mortality from physical locations that themselves are perfectly safe for the viewer. The storage of the Dead Sea scrolls in escarpment caves in eastern Israel best proves this point (Figure 45). The priests who transcribed these documents selected sites that would be absolutely safe for the scrolls and absolutely dangerous for any creature hunting them.

Cliffs thus performed a dual function for early humans. They were a source of predictable safety and comfort, but at the same time they could kill animals including humans. We argue that cliffs and their caves represent sites with an incredible and immediate juxtaposition of human emotions and feelings of both insecurity and security. A tremendous psychological tension develops between the absolute impermanence of life

Figure 45.
The site of the Dead Sea Scrolls at Qumran, was apparently selected
because the rock outcrops provided: 1. environmental conditions that
would protect the scrolls for long periods of time and
2. difficult access that would restrict vandals. Photo: J. Lundholm.

and the absolute permanence of the security of caves within escarpments. An immediate compression of security and fear is felt when one visits an escarpment, whether at the base or the top. We can think of no other habitat type or landform that has this effect on people, and we conclude that these feelings have given rise to the reiteration of cliff-like images in so many areas of human existence. René Dubos (1965, 1968) argues that humans, like all organisms, must retain structural and perhaps even psychological records from their evolutionary past. He says that refuge habitats offering prospect and access to water at the same time would have been exceptionally valued. Many others agree with this interpretation.[279] They argue that humans today should reconstruct such places to reinforce these strong feelings and attitudes with evolutionary origins. Dubos adds that such places included rocks, springs and river courses. Spirn (1998) contends that landscape architects should acknowledge and exploit these feelings to create a more comfortable and productive working and living environment for everyone.

Rock as "home"

The fundamental attraction humans have to cliffs, rock outcrops, and caves is based on the ecological functions that these places have allowed us to carry out for a time period that only ended about 10,000 years ago. Until that time, rock outcrops and cliffs were nurturing places. Tuan (1974) notes that our ancient human ancestors had

> moved out of the forest to the plains [where] they sought the physical (and one might guess) psychological security of the cave. Artificial shelters are man-made concavities in which life processes might function away from exposure to light and to the threats of the natural environment. The earliest constructed dwellings were often subterranean.

All organized religions date more recently than this time period. We can argue, therefore, that even for the oldest known current faith system (possibly Judaism or Hinduism) or the oldest that ever existed (polytheistic Egyptian or Greek), the central role played by rocks, escarpments, mountains, caves, cliffs, or their built substitutes was one that had existed prior to the historical period by several hundred thousand and perhaps even one million years. Rock outcrops, cliffs, and the caves they contained represented centrally important habitat features that provided humans with critically important and limited resources unobtainable elsewhere. These habitats allowed humans to persist during various episodes of glacial advance and retreat over the Pleistocene. Such a long period of time represents over 30,000 generations of humans (assuming a modern 30-year generation time). This easily represents enough time for natural selection to operate on human brains to reinforce the ecological values of rock outcrops and cliffs by generating feelings of instinctual attraction to these places. This attraction, we argue, has been forgotten by most of us, even while the fundamental attractiveness of the rocks has been ritualistically and systematically incorporated in the world's major religions or faith systems. We also believe that we have subconsciously incorporated different aspects of the structure and function of cliffs, caves, and rock outcrops into the buildings that we currently manufacture for our work, pleasure, and spiritual activity.

Leaping to the present, we believe that urban environments may reflect nothing more than the same primitive attraction to habitats that at one time offered prospect and refuge. The current attraction that modern humans have to cliffs, rock outcrops, steep slopes, houses, apartment buildings, and cities may be based on a kind of residual biological memory of a habitat that nurtured us sporadically and in some cases continuously for this long period of time. Further, if one revisits many of the architecturally important sites around the world, the connections to the Urban Cliff Hypothesis will begin to resonate with meaning.

225 Tilley (1994), Kostof (1985).
226 Guidoni (1978).
227 McNatt (1996).
228 Courbon et al. (1989).
229 Shippee (1934).
230 Routledge (1921).
231 Ault (1922).
232 Humphrey (1995).
233 James (1965).
234 Sandars (1972).
235 Several articles are relevant here including Leakey (1983), von Puttkamer (1979), Cameron (1950), Marshack (1975), Stuart (1981), Casteret (1948), Shor and Shor (1951), and Jordan (1979).
236 Southern and Saharan Africa: Solomon (1996), Coulson (1999), and Lajoux (1963); Europe: Mithen (1990); the Americas: McNatt (1996).
237 Lewis-Williams (2002).
238 Stoffle et al. (2000), Brady and Prufer (1999), and Sikkink and Choque (2000), respectively.
239 The oldest site yet recorded is at Cosquer Cave near Marseilles, France. This site was unknown until the 1990s because it was flooded by the rising sea levels of the Mediterranean during reglaciation.
240 Mellart (1964), Clottes (1995).
241 Rudofsky (1977).
242 Turner (1966). Turner invites close inspection of the Mona Lisa and other Renaissance paintings whose foregrounds are dominated by portraits but whose background is dominated by scenes of wild rocky or escarpment habitats. He argues that the image of the Mona Lisa is the complement to the primordial world of the cave.
243 Jackson et al. (1965).
244 Dams (1984, 1985), Fagg (1956), Fagg (1997), Cross (1999), and Cross et al. (2002) suggest that lithophones were constructed and used during Paleolithic times and continue to be used today, especially in Africa and Asia.
245 A gamelan is a type of orchestra common to Southeast Asia, consisting mainly of tuned metal or wooden chimes and other percussion instruments.
246 Maccoby (1982), Jacobsen (1976), Davis (1981), Hogg (1958).
247 Lewis-Williams (2002).
248 Maccoby (1982).

249 Gamble (1994).
250 However, as Gargett (1999) points out, caution should be used in interpreting the evidence. Most studies reporting Neanderthal and early modern human burials have not adequately excluded natural processes to explain the preserved human remains.
251 Larson et al (2000a).
252 Lee (1924).
253 Lerche (2000).
254 Jacobs (1968), Tilley (1994).
255 Bernbaum (1997).
256 Schaeffer (1930).
257 Mulliken (1938).
258 James (1965).
259 Groff and Lau (1937).
260 Baring-Gould (1911).
261 Perkins (1909).
262 Clark (1977).
263 Howell (1987), Tilley (1994).
264 Abdul-Rauf (1978).
265 Jacobs (1968).
266 Bernbaum (1997).
267 Alexander (1977).
268 Tilley et al. (2000), Casey (1993).
269 Tilley (1994).
270 Kellert (1993), Kellert and Wilson (1993).
271 Appleton (1975), Zube et al. (1975), Ribe (1989), Hull and Stewart (1992), Heerwagen and Orians (1993).
272 Kaplan (1992)
273 Balling and Falk (1982), Ulrich (1993).
274 Larson et al. (2000a).
275 Orians (1986), Orians and Heerwagen (1992).
276 Pringle (1988).
277 Gamble (1994).
278 Wentzel (1953).
279 Morrish (1996), Tuan (1974), Hirsch and O'Hanlon (1995).

CHAPTER   SIX

Ancient Cities, Modern Caves: Architecture and Urban Landscapes

We believe the story we have tried to tell thus far in this book has the potential to change our understanding of the environment we have created for ourselves. As mentioned at the beginning of the book, the Urban Cliff Hypothesis has implications for architecture, landscape architecture, nature interpretation, and ecological restoration. Those implications are explored in this chapter. First, we will show how the choices we make in designing our modern surroundings may still be subconsciously guided by our cliff origins. Then we will argue that the Urban Cliff Hypothesis provides an alternate point of view that can help us cope with the changes our success has brought to our natural environment.

Cliffs, caves, and architecture

The study of architecture stretches back more than a thousand years, and every conceivable interpretation has been put forward concerning the significance of the various forms of construction. It is commonly assumed that the spaces we create for ourselves to work and live in reflect only what we currently need or can afford. Few have considered whether our cultural and evolutionary history might influence the form of our built structures by determining which configurations of space appeal to us more than others. We propose here that we have been unconsciously influenced by an affiliation for rocky habitats in the way we design and construct our modern-day buildings, and that this has resulted in a built environment that in many ways duplicates the characteristics of our original dwelling sites.

We believe that several structural features of cliffs and caves have been incorporated into present-day architecture. There are also several features that have not been, but *could be* incorporated. We will argue that recognizing these structural features for what they represent will make it possible to enrich current architectural design with spaces and structures borrowed directly from our rock outcrop "homes."

The rampart

All rock outcrops and cliffs with habitable caves are underlain by a rocky pediment. The gradual erosion of the rock face results in the accumulation of rocky debris called talus or scree at its base, which adopts a fairly predictable angle of repose ranging between 34 and 40 degrees. (No such debris slope accumulates if a river or lake margin comes directly to the bottom of the cliff. However, since there rarely is habitable space at the base of the cliff in these situations, we need not consider this further.) Because of this, the approach to all caves involves an upward traverse across a field of open rock. People approaching the cave are therefore at a disadvantage with respect to those already at the cave: the upward traverse requires more physical exertion than does a downward traverse, and it is far easier to attack a person from above than below. Hence, the talus slope represents a protective device of its own.[280]

Early in the history of built structures, and particularly in the construction of fortresses or religious sites,[281] sloping ground was intentionally incorporated into the access points to the buildings. Medieval forts and castles in Europe, ziggurats in Mesopotamia, and pyramids in Egypt and Central America are all characterized by large expanses of sloping terraced ground at the entranceway. Morrish and others[282] argue that most of these stone monuments were designed to be built replicas of natural rock outcrops, cliffs, or mesas. At Ur in Mesopotamia (Figure 46), where the terrain was lacking relief entirely, the topographic profile was accentuated by constructing three layers totalling 70 feet (22 metres) of absolute elevation to form the Ziggurat.[283] Most prominent monuments in major cities of the world have multiple arrays of staircases that sweep upward from the roadways providing pedestrian access. An example is the Capitol building in Washington, D.C. (Figure 47), where several series of steps must be traversed before the stone building is entered. Such ramparts leading to the access points of forts, cathedrals, monuments, and official buildings place those ascending to the entrance at a psychological disadvantage.[284] In contrast, the view from the top of the rampart on exiting the building is one of authority and dominion over those below the entranceway, engendering a feeling of power.

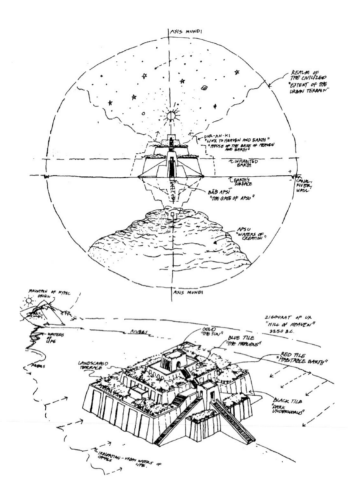

Drawing 18.
THE BRICK MOUNTAIN

Figure 46.
The Ziggurat of Ur and other religious buildings in Mesopotamia were
designed to recapitulate the mountain homes of the peoples who first settled
in the river valleys of the Tigris and Euphrates Rivers. Illustration taken from
Morrish (1996). Used with permission.

Figure 47.
Important government and religious buildings commonly present a large
array of carved stone steps leading to the entranceway. Such approaches
recapitulate the talus slopes leading up to the mouths of caves.
Photo: P. Kelly.

Orientation

The orientation of a cliff or escarpment controls the energy budget and hence the physical environment of the dwellings within it. This can be taken into consideration when selecting caves for habitation.[285] In cold climates of the northern hemisphere, south-facing cliffs receive the maximum radiant exposure in mid-winter. In hot climates, on the other hand, a north-facing exposure minimizes direct solar radiation.[286] Seasonally varying wind speed and direction can also be taken into account. Exactly the same considerations have for the past 30 years been incorporated into the design of new buildings. Interest in passive heating and cooling first emerged during the oil embargos of the 1970s and has increased since then because such buildings are more comfortable as well as cheaper to operate.

The orientation of the cave entrance also has a direct bearing on the control of competitors and predators. The preferred orientation would either restrict access to the cave by potential attackers or provide a prospect view of the possible approaches to the entrance. The same principle is applied to the design of houses, apartment complexes, civic buildings, churches, monuments, and forts. Architects rarely position the openings to such buildings in a way that obscures the prospect view from the entranceway. Buildings that allow residents to scan the horizon for potential threats are preferred over those with a limited lateral view of the surrounding terrain.

The facade

The appearance of the front of a cliff face or building has a strong influence on how the dwellings within it are valued.[287] The sites of famous cave dwellings such as Niaux, Lascaux, Les Eyzies, and Chauvet in France present an awesome spectacle to human observers, whereas caves in smaller escarpments that are concealed from view appear to have been used less consistently. Qualities that appear to be important in terms of overall appearance of the facade include height, length, colour of the surface, heterogeneity of texture, and luxury of sheathing vegetation.

All of these items have at times and to varying degrees been incorporated into built structures. Building height has always been a hallmark of societal importance, and it is well established that work or living spaces on the top level of buildings (the penthouse) are considered more desirable than spaces at intermediate levels. The imagery of "rising to the top" is pervasive in the corporate sphere, and this idea is well exploited in commercial advertising.

The Hanging Gardens of Babylon perhaps represent the best-known and most ancient example of this practice, but architectural planning since that time has integrated living organisms such as trees, shrubs, vines, and rock gardens into the design of the entrances and surroundings of buildings. This is appreciated by people who use the spaces,[288] and there are some modern examples, such as the Trump Tower in New York City, where architects have incorporated ledges as tree planting sites on the front facade of the building.[289] Raptors such as the peregrine falcon have been successfully reintroduced into modern cities where they build eyries in various parts of the facades of large buildings, but architects have not consciously incorporated structures that provide nesting sites on these surfaces.

The drip line and the foyer
Inhabited caves almost always occur in sedimentary rocks that erode more rapidly at the base of the rock face than at the top. As a result of this undercutting, no direct precipitation falls under windless conditions within a zone at the cave entrance that is bordered by the drip line.[290] People value living spaces more when the drip lines are large and effective. The effectiveness of the drip line depends to a large degree on its orientation. For example, a space that is protected from the prevailing wind is much more pleasant than one that faces into it.

The space between the drip line and the deeper, only dimly lighted parts of a cave is optimally suited for activities requiring both light and shelter from rainfall. This area was most likely the effective daytime living space of the inhabitants as well as the place where communal work activities were carried out and meals were taken. According to some reports,[291] most modern human sites in Great Britain had hearths located just inside the drip lines of the caves. If the fires had been set more deeply in the recesses of the cave, smoke inhalation would have probably represented a serious health risk. Hearths positioned near the mouth of the cave also made it possible for activities requiring light to be carried out in the comfort of radiant heat from fires. Food preparation taking place near the hearth would have resulted in an accidental tendency for meat and plant material to be preserved by smoking.

Most dwellings built by humans incorporate some kind of drip-line structure into their design (Figure 44). The area between the drip line and the interior living space commonly consists of an outside dry but otherwise unregulated space called porch or deck that continues into a climate-controlled space behind doors referred to as foyer, lobby, or vestibule. The direct correspondence between these features of modern architecture and our ancestral caves is still evident from the Latin root of the word *foyer*, which means *hearth* or *fireplace*.

Communal or semi-communal activities still take place in or near the lobbies of modern-day buildings. Many hotels have small kiosks, boutiques, or shops facing the lobby, and dining rooms, lounges, or meeting rooms adjacent to it. Apartment buildings may have communal laundry areas or party rooms that are located between the public areas at the entrance and the private living space in the interior.

Resource storage

All natural cave dwellings must have had places where food, water, or other resources were stored. Little direct evidence remains of these places, but pollen, carbonized seeds, and bone fragments found in cave deposits show what types of food were consumed and how they may have been processed and stored. The placement of these storage areas near the communal meeting places ensured that the resources were protected from pests such as rats, mice, pigeons, and cockroaches, as well as from other humans.

Modern humans appreciate access to resources both in communal spaces and within each individual living or working unit. Present-day architecture has taken this into account by providing a balance of resource storage and use areas that matches the expectations of the people using the structures.[292] The coffee bars, vending machines, and water fountains found in lobbies and the refrigerators, pantries, closets, and lockers in individual apartments are all equivalent to the crevices, ledges, and subchambers that our ancestors used within caves and rock outcrops.

Sleeping areas

We do not know the precise locations where ancient hominids slept, since this activity leaves no evidence behind. Anyone who has attempted to sleep in a cave, however, will attest to the unpleasantness of a rough rocky surface. Noise, light, and smoke near the cave mouth would have made recesses within caves the most likely places used for sleeping. The complete segregation of individuals or family units from each other was probably impossible in small caves. Built structures would have significantly improved the opportunities for people to seek shelter from each other as well as from sound, light, and smoke.

The minimum dimensions of sleeping units are controlled by the size of the human body. Most modern-day architects design sleeping spaces to be large enough to permit standing and walking around, but certain modern Japanese hotels have, for cost reasons, minimized them to the point where an individual cannot stand upright in them. These miniature sleeping chambers and the sleeping units used on transcontinental railway trains probably reflect best the sleeping-berth type arrangements present in many rock shelters and caves.

Living space for animals

As we discussed in Chapter 2, cave dwellings sheltered not only humans but also a variety of animals. Among these were undesirable pests such as rats, mice, pigeons, and insects, but also mutualists such as dogs and cats whose presence was encouraged because of the beneficial services they provided (scavenging of waste, rodent control). Livestock, on the other hand, never shared the caves with humans, as is clearly shown by the absence of their bones from cave deposits. The reasons for this may include the defecation habits of these animals as well as their large size. It was only with the appearance of built structures that special facilities were included to house animal species such as cattle, horses, goats, sheep, and pigs.

The same principles still apply to our modern-day living arrangements. Cats and dogs are considered to be normal co-occupants of buildings both large and small and are generally welcomed to share our dwellings because they provide us with psychological or physical protection and comfort. Livestock is still housed in special facilities that are separate from the human living space. Modern buildings tend to be well sealed, well lit, and relatively free of accumulations of waste and organic debris. This may reduce their attractiveness to commensal organisms, but rodents, birds and insects still exploit human-built structures for nesting, foraging, and food storage, despite our efforts to discourage them.

Latrines

Unlike many animals, humans have a sensory aversion to their own waste and have always used dedicated latrines to avoid defecation and urination within their living space.[293] This represents an effective mechanism for the avoidance of pathogens. It is thought that the latrines associated with cave dwellings were located outside the cave mouths but not too far from the dwelling. Likewise, most present-day cultures emphasize sanitary disposal of human and animal waste, and latrines are built both within private living spaces and in communal locations.

Niches for art and spiritual connections

The deepest recesses within caves were generally reserved for spiritual and artistic purposes. In our view, the special rooms for contemplation and spiritual connection that are often included in modern building design are nothing but copies of similar spaces created by early humans living in rock shelters or caves. Like caves, modern built structures support art on vertical surfaces and occasionally on ceilings. The cliffs that support rock art, for example in Bhutan (Figure 48), serve many of the same functions as the walls of a modern art gallery (Figure 49).

The importance of art on walls and ceilings to early and modern humans cannot be overemphasized. Our standard of living is commonly measured in terms of the total cash value of accumulated assets, but the standard of living that each person *feels* must represent a collective sum of all that they have accomplished or accumulated throughout their lives. An appreciation of "self" in the context of what others have accomplished is

Figure 48.
Cave and cliff art is found around the world. Food plants and animals, as well as ritual ceremonies, are often displayed. These locations inhibit the deterioration of the images by protecting them from physical weathering.
Photo: L. Melville.

important. However, it is often difficult to recall all of what one has done and felt and to mentally reconstruct places or people that have been visited. Graphic art provides the opportunity to relive the events in our own lives and to imagine and enjoy those in the lives of others. Such art supplies each of us with connections to worlds that once existed and are now gone, or that still exist in places that are inaccessible to us. We believe that much of the graphic arts today, including still and moving images, television, film, and video, provide exactly the same function to modern humans that the rock art on the wall of the Agawa Canyon in Ontario, Canada, or the walls of the caves of Lascaux, France, provided to prehistoric humans. Modern architecture is most effective at binding people to the artificial enclosed spaces it creates when it gives them the opportunity to draw experiences from this art.

Modern architectural design, then, subconsciously mirrors many of the features that were present in habitable caves and rock shelters. There were limited options for modifying and improving the design of cave dwellings, but the invention of the brick, and later of mortar, allowed the elimination of most or all of the defects of natural shelters. However, the Urban Cliff Hypothesis suggests that humans have subconsciously retained and

Figure 49.
The Metropolitan Museum of Art in New York City represents the modern-day version of a cavern filled with cave art. Photo: P. Kelly.

ANCIENT CITIES, MODERN CAVES

incorporated in modern structures many of the positive features of our ancestral homes, features that have contributed to making us productive and successful as a species over the past million years. The implications are obvious for architects, who could use this knowledge to make the design, appearance, function, and location of buildings more comfortable and productive for the humans and animal species housed in them.

The origins of landscape architecture

The Urban Cliff Hypothesis also provides insights into the reasons why humans feel strongly about the physical layout of the environment external to their dwellings.[294] The surroundings of a dwelling make a statement to the residents about themselves, and a statement to others about who the residents are and what they value. This has been shown in the case of Medieval castle gardens, which were designed to communicate to visitors the relative wealth and power of their owner.[295]

Contemporary houses, both single-family and multiple-family (Figures 50, 51), and the landscapes surrounding them still reflect many of the features of our original cliff and talus slope habitat.[296] The rock-faced buildings contain a wide variety of sub-spaces, each of which has a different function. The entranceway is at the top of a slight incline cut into rock, and is protected by a drip line. The fronts of the buildings are often landscaped with species that thrive in broken talus, including tulips and eastern white cedar. Even one of our original companion species, the cat, is still evident. These houses truly are their inhabitants' castles.

These design principles appear to be well understood by students of landscape architecture, but the ecological basis for them has been little explored. The Urban Cliff Hypothesis suggests that the persisting urge in humans to create upwardly sloping entranceways to their dwellings and surround them with productive gardens is fully explained by the habitat history of our species. The Old Persian word for a rock-walled garden setting is *pairidaēza*: from this word is derived the English word *paradise*.

As we will emphasize in the final chapter, these gardens at the entrances to dwellings and other buildings therefore represent paradise in the original and literal sense of the word. Simonds and others[297] have argued that landscape architecture should borrow from nature and use the built structure to protect and nurture us. Thus the entire discipline and practice of landscape architecture can be viewed as the art of constructing paradises for people on a variety of scales.

Figure 50.
A montage of four separate single-family dwellings in Guelph, Ontario. Each is characterized by a slight incline to the front door, an overhanging roof, a semi-protected space beneath the drip line, and rocky gardens enriched in cliff or talus slope species. Photos: P. Kelly.

Figure 51.
Habitat 67 in Montreal, Quebec. This dwelling was constructed in 1967 and remains as an example of modern organic architecture that blends vegetation and concrete into the living environment. Photo: P. Kelly.

Planet of cliffs

The Urban Cliff Hypothesis should profoundly change the way we think of natural cliff, rock outcrop, and talus slope habitats. Instead of being considered marginal and rare with limited relevance to society as a whole, cliffs should become examples of central and ubiquitous habitats and the ancestral homelands that may have given rise to many of the plants and animals associated with us. This idea leads quite naturally to the topic of global biotic homogenization discussed by Quammen (1998).

The central idea behind the Global Biotic Homogenization Hypothesis is that human disturbance has created opportunities for species with broad ecological tolerances. Such species are successful in human-disturbed environments at the expense of native vegetation and wildlife that are more specialized in their ecological requirements. Quammen and others believe that this displacement represents the next major mass extinction on the planet and argue that it should be stopped.

Let us start off by agreeing with Quammen that the flora and fauna native to an area are the most desirable functionally and aesthetically. But since human disturbance is likely to be a permanent presence on the planet, it might be appropriate to have a closer look at the species that benefit from it. Quammen lists rats, lice, several microbes, burr plants, dogs, pigs, goats, cats, cows, house sparrows, starlings, pigeons, house geckos, ragweed, cockroaches, and a wide variety of other commensal or domesticated creatures as ones that are doing particularly well worldwide at the expense of others. Readers will immediately recognize many of these plants and animals as organisms listed in Table 1: species associated with humans whose habitats of origin are cliffs, rock outcrops or other persistent rocky waste places. Quammen is concerned about species that are either exploited by humans in agriculture or are commensal with humans as pests and weeds. As we showed in Chapter 2, most of these species have a common origin: they were our neighbours on our ancestral cliffs. Originally harvested as wild species, some of them were subsequently domesticated while others took advantage of our destructive behaviour and joined the human exodus from cliffs to exploit the open, disturbed, often highly fertile environments associated with the human presence all over the world. Plant species such as the ribbonfern (*Pteris cretica*), the bay laurel (*Laurus nobilis*), the sea-carrot (*Daucus carota*), and New Zealand flax (*Phormium tenax*), which still occur on natural rock outcrops or cliffs, in open waste places, as well as in cultivated places,[298] demonstrate the ecological connection between these apparently unrelated habitat types.

We can thus extend the Global Biotic Homogenization Hypothesis by concluding that opportunistic species originally from open rocky habitats are the ones succeeding at the expense of native species adapted to environments that are neither open nor disturbed.

However, we feel that we should not vilify the array of weedy plants and animals on the planet, and we do not agree with Quammen that the "Planet of Weeds" scenario is necessarily bleak. The Urban Cliff Hypothesis *predicts* that species once restricted to rocky habitats will follow the trail of human disturbance to fill the vacancies created by the removal of the native flora and fauna (Figure 52). The fact that the most abundant animals in the world are exceedingly difficult to eradicate may not be accidental. In the United States, $136 billion is allocated annually to the

Figure 52.
A bridge abutment in Guelph, Ontario, supporting spontaneous populations of weeds that are natural rock outcrop endemics. Dandelion, wild carrot, hawkweed, and similar species are common in such settings.
Photo: D. Larson.

control of non-indigenous species and $70 billion alone is spent on weed plants, rats, mice, cats, dogs, and pigeons.[299] The high cost of removing these organisms from our cities may simply reflect the exceptionally high level of ecological stability or persistence that these rock outcrop species have within our built environment. We argue that since it has proved difficult or even impossible to stop the human race from disturbing the planet to create opportunities for themselves, we should consider ourselves lucky that at least *some* plants and animals can tolerate and even exploit the conditions we create.

These weedy, opportunistic invaders become far more attractive when viewed from the perspective of cliff ecology. We are far from suggesting that native flora and fauna be destroyed to benefit cliff or talus slope species, but we do contend that this collection of species is better than none at all. Quammen argues that humans have destroyed much of the planet's biodiversity creating opportunities for ugly opportunistic weedy species. The Urban Cliff Hypothesis provides the counterargument that humans have indeed transformed the planet, but into a habitat that resembles our ancestral home and that contains real plants and animals that still reflect the habitats where they evolved. Unlike Quammen, we respect the organisms that are taking advantage of human benevolence by expanding into the ideal habitats we keep creating for them.

Restoring paradise

How we regard the former cliff species that are now associated with humans has important implications for ecological restoration, a research area in which scientists and land managers try to reconstruct a deteriorated landscape to closely resemble the original landscape present before the disturbance.[300] Deciding on and defining the exact form of the desired result, referred to as the *restoration target*, has proven to be one of the most difficult aspects of restoration ecology, and there has been much recent but fair criticism that restoration ecologists were carrying out large-scale gardening because of a failure to clearly define the restoration targets.

Restoration targets are reasonably well known for many different types of habitats, including tall and short grass prairie, Sonoran desert, tropical forest, temperate deciduous forest, and salt marsh. However, situations may arise in which enormous amounts of expensive environmental remediation would be required before restoration to the original state could proceed. In principle, it is preferable for the restoration target to be the original flora and fauna of a site, but this may be impractical, and there may be opportunities to restore disturbed landscapes using targets that never previously existed at those sites. Abandoned factory lots in city cores, for example, may resemble limestone or shale barrens in their chemical and physical conditions,[301] and may already be colonized by weedy, alien, or perhaps even invasive plants and animals. The reason why these particular organisms are invading is that they are accustomed to such sites, that is, they find in them some or most of the habitat conditions that they experienced during their evolutionary history. By recognizing that every region of the world will have some *native* species with the same habitat requirements as the opportunistic aliens, it may be possible for people to intervene and actively replace the aliens with native species that fulfill the same ecological functions within the context of the native biodiversity. Spirn (1984) has shown that many urban environments contain habitats that are concrete, glass, and steel manifestations of naturally occurring habitat types. This idea should be familiar to most people who readily accept the idea that cities such as New York are genuinely concrete canyons.[302] We argue that the spontaneous assemblages of plants and animals that occur in these urban canyons can be modified directly to result in true native communities that are equally well suited to the conditions as the exotics. If allowed to develop on their own, some of these communities can be as beautiful and enjoyable as any natural habitat, as illustrated by the history of the Colosseum in Rome.

The restoration of the Colosseum

Most tourists experience the Colosseum in Rome as a beautifully preserved building that represents the peak of architectural development in the Roman Empire (Figure 53). The building is evidently secured against further deterioration by a large team of masons and architects. What today's tourists do not appreciate is what was lost to the world when the architectural restoration of the building was begun in 1871. In his chronicles of the building, Quennell (1971) includes quotations from people who viewed and studied the Colosseum prior to its restoration and reported that a lush native vegetation had encroached on the structure. Surveys of the flora of the building conducted in the mid-1800s[303] found 421 species of plants growing there. These include many that are native to rock outcrops and escarpments, for example capers, olive, fig, grape, cyclamen, and saxifrage. Quennell quotes Shelly as having written in 1818:

[The Colosseum] has been changed by time into an amphitheatre of rocky hills overgrown by the wild olive, the myrtle, and the fig tree, and threaded by the little paths, which wind among its ruined stairs and immeasurable galleries: the copsewood overshadows you as you wander through its labyrinths, and the wild weeds of this climate of flowers bloom under your feet. The arena is covered with grasses, and pierces, like the skirts of a natural plain, the chasms of the broken arches around.

He also quotes Thomas Cole, who in 1932 wrote:

From the broad area within, it rises around, arch above arch, broken and desolate, and mantled in many parts with laurustinus, the acanthus, and numerous other plants and flowers . . . It looks more like a work of nature than of man . . . The regularity of art is lost in dilapidation . . . Crag rises over crag; green and breezy summits mount into the sky.

Quennell then himself adds:

Many trees flourished on lofty ledges – the fig, the cherry, the pear, and the elm. The ruins also supported vines and ivy, clematis and wild roses; and the stonework was pied and dappled with an exquisite variety of small plants, which included rosemary, thyme, sage, cyclamen, daisies, pimpernels, hyacinths, saxifrage, violets, strawberries, marigolds and larkspur.

These descriptions clearly show that after about 500 A.D., the deterioration of the Colosseum resulted in a progressive recolonization of the site by native plants and probably animals as well. This ecological succession produced the wondrous and beautiful habitats described in the above quotations. We can observe the same process of natural recolonization today on ancient buildings in many countries of Europe: for example, Figure 54 shows the walls of a deteriorating castle in the United Kingdom that are luxuriantly vegetated with *Taxus baccata, Taraxacum officinale, Polypodium vulgare, Asplenium ruta-muraria, Asplenium trichomanes,* and *Cymbalaria muralis.* On the Colosseum, of course, the natural succession came to a halt in 1871 when the physical restoration of the building began, and many old buildings throughout Europe are "sanitized" at regular intervals to remove any vegetation that has established itself.

The important point is that the assemblage of species that ecological processes had recruited on the Colosseum walls over a period of 1300 years are those that recolonize the rock walls left behind in abandoned quarries,[304] and the same that normally grow on natural rock outcrops and escarpments. Returning to the topic of modern restoration ecology, we believe that the natural revegetation of both quarry walls and abandoned building sites can probably be accelerated by the selective introduction of native plants and perhaps animals. This would help increase the biodiversity of the region and allow the ecological connections to form quickly and solidly. Moreover, as the example of the Colosseum has shown, these places can become beautiful over time. Adopting this new approach will require a switch in thinking in restoration ecology regarding restoration targets. We are advocating that we should not try to reconfigure the habitat to make it suitable for the species that occupied it *prior* to the disturbance. Instead, we should consider, for use in a restoration protocol, the native flora and fauna that can recognize and exploit a site *as it stands.* Rather than trying to revegetate abandoned building sites with the forest

Figure 53.
The Colosseum in Rome has been restored and cleared of its encroaching vegetation since the end of the 19th century. What is presented to the viewer now is much fewer than the verdant display of 421 species of plants that were reported from the ruins in the mid-1800s. Photo: D. Larson.

Figure 54.
The walls of medieval castles in the United Kingdom support a luxuriant
flora of trees, herbs, and ferns, including yew (*Taxus baccata*), dandelion
(*Taraxacum officinale*), Kenilworth-ivy (*Cymbalaria muralis*), rock polypody
(*Polypodium vulgare*), spleenwort (*Asplenium trichomanes*), and wall rue
(*Asplenium ruta-muraria*). Photo: P. Kelly.

that was there before the human disturbance, we should use the species
more suited: those native to rock outcrops and cliffs. Restoration to a
suitable target is much less expensive than conventional restoration to the
original target. The original target might be preferable in the long run, but
this disadvantage is offset by the ability to carry out many more such
restorations with much less money and in a shorter period of time, since
little more may need to be done than make available the propagules of
native species. We have seen[305] that abandoned limestone quarry sites
return to a species composition that resembles natural limestone cliff faces

within 70 years. Observers had noted[306] in the United Kingdom, too, that abandoned quarries assemble a diverse natural flora and fauna that bear no relationship to the forest habitats that existed prior to the initiation of quarrying. The Urban Cliff Hypothesis provides another argument in favour of this approach. We have shown above that rock outcrop, cave, and cliff habitats have long been associated with human evolution and culture. It follows that if urban or suburban structures have been built on land that once supported lush forest, but are then abandoned, then it is reasonable to convert the abandoned land into ecosystems that grow from rocks, caves and cliffs as restoration targets rather than forest, depending on the cost and difficulty of recreating forest habitat.

Purists may immediately reject this alternative paradigm on the basis that it provides a political and economic incentive to disturb and subsequently *not* restore a landscape element. We understand this argument, but counter that global biotic homogenization is taking place relentlessly and unless unlimited funds are available to return all disturbed habitats to their original conditions, resources may be better allocated to restoring lands to a self-sustaining and natural-looking target already close to the disturbed lands. Much land that is considered too expensive to restore to its original vegetation might be cheaply converted into communities resembling those found in rock outcrop habitats. This type of restoration could occur side-by-side with the restoration of the highest quality remnants of other habitat types.

Regardless of the philosophical and political implications of this view, at the very least we can acknowledge that many highly disturbed areas that are not even considered for restoration, building walls and pavement cracks for example, could actually support native vegetation. The spatial extent of restoration in urban areas could thus be greatly increased. New research in restoration ecology is required to investigate the suitability of abandoned buildings as sites of refuge for native biodiversity as well as the ability of native organisms to grow and perhaps even overcome alien organisms in contaminated sites. Unlike the natural recolonization of Roman buildings during the past 1500 years, revegetation dynamics in our industrial age must take place within the context of air, soil, and water pollution as well as physical disturbance. The array of microorganisms, plants, and animals that exist on derelict industrial and agricultural sites around the world may, as Quammen concludes, represent those few organisms that can tolerate physical and chemical contamination at the same time.

280   Maass et al. (2000).
281   Toy (1985).
282   Morrish (1996), Beek (1962).
283   Tuan (1974).
284   Orians and Heerwagen (1992).
285   O'Sullivan (1994).
286   This is of course reversed in the southern hemisphere.
287   Maass et al. (2000).
288   Im (1984).
289   Cooper and Taylor (1996, 2000).
290   Larson et al. (2000a).
291   Smith (1992).
292   Alexander (1977).
293   Nesse and Williams (1996).
294   Cooper and Taylor (1996).
295   De Turk (1968).
296   While we have emphasized the types of houses used in industrialized, temperate-zone cultures, the same remark likely applies to the rich diversity of permanent dwellings used by contemporary humans in a wide variety of cultures.
297   Simonds (1961), Laurie (1975), Cooper and Taylor (1996).
298   Stace (1999).
299   Pimental et al. (2000).
300   Jordan et al. (1987).
301   Anderson et al. (1999).
302   Simpich (1930).
303   Sebastiani (1813) and Deakin (1855) quoted in Quennell (1971).
304   Ursic et al. (1997).
305   Ursic et al. (1997).
306   Usher (1979).

CHAPTER SEVEN

Paradise Found, Forgotten, and Rediscovered

In this book we have tried to replace a series of conventional views on separate topics (such as human evolution, origins of agriculture, and architecture) with one large story that binds these ideas together. This story can now be told through a series of drawings in 12 panels, starting with Figure 55a and ending with Figure 55l.

The storyline is quite simple. There was a paradise once, a rock-walled garden. It was inhabited by plants and animals that were not very aggressive or assertive when competing with other organisms, and their survival was dependent on the refuge provided by the rock (Figure 55a, b). What the organisms did have, however, was the ability to take advantage of sudden changes in resource levels: when conditions temporarily improved, they could grow like weeds. These plants and animals were not the generalist species that most people think. They were exposed to and adapted to harsh physical conditions and to resources that varied enormously over space and time. They were tough, opportunistic species that could patiently wait out bad times and then grow explosively when conditions improved. People use the word *generalist* to imply an ability to put up with anything. But if one thinks about this, being extremely adaptable over time and space is itself an incredible specialty.

Human ancestors eventually found and started to exploit this rocky paradise when climate change forced them out of the forests and into the savannas where they were ill-adapted to coping with the daily climatic extremes (Figure 55c). Like many other organisms using the rock, humans were relatively defenceless against large predators. But the resemblance did not end there. As it turned out, humans were almost as opportunistic as the plants and animals already occupying these habitats. For perhaps 500,000 years, humans continued to live in the rocky paradise in association with these other organisms. Eventually they started to use and then control fire, which allowed them to exclude some of the more dangerous competitors for caves (Figure 55d). Then came another 460,000 years where humans made use of the resources broadly available in the surrounding savannas and riverbank forests while living in caves at the bases of cliffs. Except for the control of fire and the ability to manufacture stone tools, the manufacture of items to protect people from the physical or biological threats from nature had not developed substantially during this long period of time.

At the end of this 960,000-year period, we had formed strong ecological bonds with a wide variety of species that exploited us as much as we exploited them. Whether we liked it or not, they became our constant companions, and gradually acquired ecological characteristics that made them increasingly dependent on us (Figure 55e). Between 30,000 and 12,000 years ago we began to modify the physical space of the rock shelters (Figure 55f) and eventually to construct our own shelters in open terrain. Since the building materials used were available locally, the built shelters could be located far from natural escarpments (Figure 55 g). As we began our exodus to the built environment, some of our companion plants and animals remained behind on the natural cliffs. Others, however, stayed close on our heels since we had re-created their prime habitats (Figure 55h). With the use of more permanent building materials, segregated spaces that allowed food plants and animals to be separated from people soon became a possibility (Figure 55i), and the development of substantial architecture began about 9000 years ago (Figure 55j). The built environment also provided spaces for our unwanted followers to exploit. Slowly, the conscious memory of the rock outcrop as paradise was lost or forgotten, but, as we argue, remained buried in our collective subconscious. As a result, certain configurations of the built environment have always made us feel comfortable and safe. Huts, houses, castles, and skyscraping towers all have the feel of home because each of them has features of the original paradise. All of them offer prospect and refuge at the same time, security from climate, competitors, and predators, a place to grow food and deposit waste, and a place to communicate ideas of culture and society. Our modern cities are dominated by individual structures that repeat this theme at various scales (Figures 55 k, l). At the largest scale, one could consider the entire urban landscape as a place providing refuge and prospect view at the same time. The idea of cities as safe homes might help explain why people around the world continue to flock there when economic difficulties arise.

Here, people begin each morning by rising from their sleeping chambers and getting prepared for the day's routine. Their rooms are positioned behind solid walls manufactured to be a protective barrier from threatening creatures or from human competitors. The initial daily activity, of course, is disturbing to parasites and commensal organisms, and the mice and cockroaches go scurrying for cover while the bedbugs have already retreated. The people use water for drinking and cleaning and consume the first meal of wheat, rice, or corn in a different chamber, one that contains a hearth whose exhaust gases are allowed to escape to the atmosphere without poisoning the residents. Sometimes this first meal includes meat from pigs, goats, cows, or sheep, as well as plants that produce sweet fruits. Dogs scavenge food scraps left unattended and cats

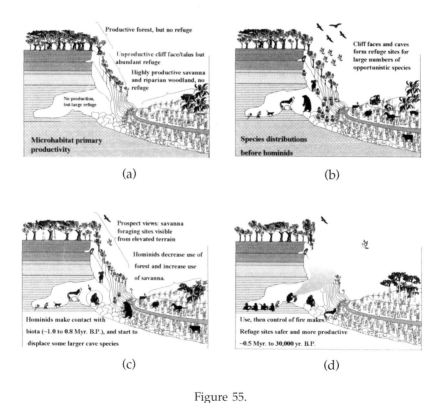

(a)

(b)

(c)

(d)

Figure 55.
Multi-frame sequence summarizing the Urban Cliff Hypothesis.
In the first four panels a–d, a series of natural scenes is portrayed in which organisms seek out rock shelters for protection and for their usefulness as sites from which predation can be carried out.

PARADISE FOUND, FORGOTTEN, AND REDISCOVERED

actively hunt the mice, rats, and pigeons. The supplies for the meal are stored in other solid chambers that are as secure as possible against competitors. Once the meal is finished the refuse is disposed of in a central area dedicated to that purpose.

Following this initial daily activity, the inhabitants leave their dwellings, walking out past the drip line into an area surrounded by plants both useful and ornamental: cabbages, carrots, rosemary, basil, apple trees, grape vines, junipers, and cedars. These species exploit enriched pockets of soil within a matrix of hard surfaces created by the buildings and paths that surround them. Leaving the gardens, the inhabitants find themselves in a maze of paths, walls, and roofs: this framework of hard surfaces divides their world into the places where they meet in small groups to plan or execute the daily ritual of training the young, foraging for new resources, or exchanging the products of their labour. These hard partitions also support a variety of microhabitats for trees, shrubs, and vines that shade

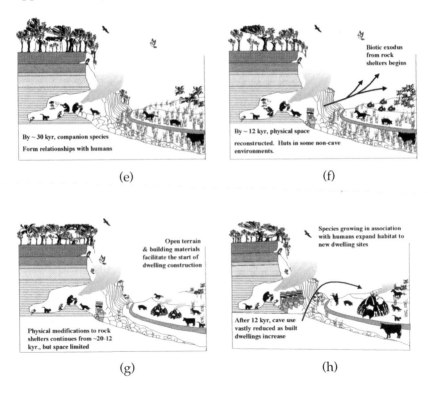

Figure 55. (continued)
In panels e–h, humans develop the capacity to modify caves
at the bases of rock outcrops so that they are more productive,
secure, and comfortable environments.

and shelter the paths and dwellings. Each of the human activities takes place in a different location: teaching takes place in areas that are well illuminated but protected from physical discomfort. People forage in more open locations where foodstuffs or water are assembled in large volume and where the diversity is the greatest.

At the end of the work day, the people retreat once more into their dwellings or public enclosures to consume the evening meal, often consisting of the same plant and animal matter as in the morning, but with more elaborate preparation and ritual. In the evening, people have the opportunity to participate in cultural activities including listening to and playing music in indoor spaces which optimize acoustics, or viewing art in other enclosures. As darkness descends, the dwellings are lit within by fire or lamps that allow a variety of productive activities to continue into the night.

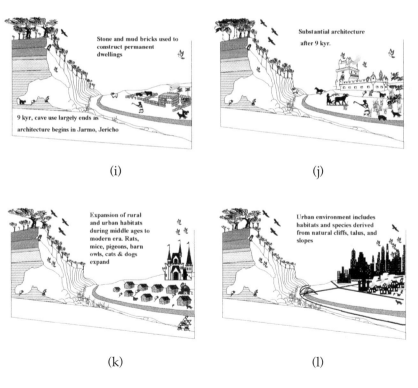

Figure 55. (continued)
In panels i–l, the exodus from rock outcrops is complete,
with humans re-creating most of the features of cave dwellings
in their built landscapes. Illustration: D. Larson.

PARADISE FOUND, FORGOTTEN, AND REDISCOVERED

The above scenario applies to people everywhere, from those living in yurts in Mongolia to those living in penthouse apartments in New York City. Not only that, but this scenario, we argue, has applied for most of the last 40,000 years and some components have persisted for about a million years.

Banishing the cave man

How would we feel about cartoons poking fun at cave or cliff dwellers if we accepted the idea that we still live in caves at the bases of cliffs, albeit cliffs that we have built ourselves? The Urban Cliff Hypothesis as presented in Figure 55 is a significant departure from conventional thinking in many fields, but when all the pieces are assembled, they make a compelling story. Some may argue that humans used cliffs or caves no more than open field sites, and we have admitted above that evidence of the occupation of open sites may have been lost over time. But if that is the case, why do people in general accept the notion of Paleolithic "cave man" as part of our past, while the equivalent labels "forest man" or "swamp man" do not exist?

A dictionary defines the "cave man" as a "prehistoric man or troglodyte who is rough and brutal, especially in his approach with women."[307] This connotation of unsophisticated backward or brutish behaviour is exactly why "cave man" conjures up an unpleasant image.[308] According to the Urban Cliff Hypothesis, modern humans are still cave men in the sense that our habitations and companion species are the very ones that we formed functional relationships with more than a million years ago. However, most people in contemporary urban societies would not take kindly to being called "cave men" or "cave women," even if they occasionally enjoy visiting the 813-year-old rock-cut Ye Olde Trip to Jerusalem pub in Nottingham, U.K. (Figure 56). In his review of the ecological history of the pigeon, Quammen (1996) says of New York City: "The human canyon. Not being descended from cliff-dwelling ancestors, I can only tolerate it for short periods myself." It would appear that, for him, even the claim that humans descended from cliff-dwellers is unpleasant to consider. Despite this, Quammen himself put forward the idea that the concrete canyons are habitats well exploited by a wide variety of opportunistic or weedy species – the same species that we have shown to include large numbers of cliff or talus slope endemics.

From our point of view, the most surprising aspect of the term *cave man* is that it makes no specific reference to the habitat setting where the caves

Figure 56.
For some, Ye Olde Trip to Jerusalem pub in Nottingham, U.K. may represent
the oldest form of refuge from the stresses of the modern world. The
structure includes rooms entirely hewn from the sandstone cliffs.
Photo: D. Larson.

were located and to the types of habitats surrounding them. The evidence
presented in the earlier chapters demonstrates that for many hundreds of
thousands of years, humans actively sought out and exploited caves *at the
bases of cliffs* – not caves accessed by holes in the ground in a horizontal
plane, but rather caves formed at the bases of vertical rock outcrops or
escarpments. According to the Urban Cliff Hypothesis, the cliffs and caves
were sanctuaries that provided shelter for people and their ecological
companions while they were exploiting neighbouring productive and
resource-rich environments for plants and animals that could be eaten. As
they developed the capacity to make bricks of mud or stone, it became
possible for them to protect these valuable food species with walled
enclosures, which is why the brick may have been the single most
important product of human creativity ever produced. Using bricks, we
started to build our own paradise, in the original sense of the word, and we
still build such paradises to this day. Modern human behaviour is no
different from that of our ancestors. Modern rock shelters in cities, towns,
and villages all over the world are used for exactly the same reasons as

PARADISE FOUND, FORGOTTEN, AND REDISCOVERED

were prehistoric rock shelters. Each of us has our own *pairidaēza* (paradise), either in a small garden plot or in a local food store. Humans make paradises all the time and everywhere without being aware of it.

Few modern humans see the resemblance between the paradises constructed today and natural cliffs. We have manufactured buildings for so long that we view them as totally artificial structures rather than reconfigurations of nature. However, from the perspective of the plants and animals exploiting it, the human-built environment functions like a fully natural habitat. At first glance, there seems to be little in common between a section of rich prairie farmland in Saskatchewan, Canada, and an open talus slope below a rocky escarpment. But the crops such as kale, wheat, barley, and oats would fail miserably to produce a bountiful crop for humans if the rich community of perennial grasses and forbs that originally grew on these prairie soils had not been utterly destroyed by humans first. Only then was it possible for the nutrient-rich humus accumulated over 10,000 years to be exploited by the crop plants that were intolerant of competition. Almost all of the crops we grow for food do not tolerate competition from other species, and yet grow very well in environments locally enriched in organic matter. Where do conditions such as these occur naturally? The only places on the surface of the earth where limited competition and an accumulation of organic debris occur *in combination* are those that are regularly disturbed by natural processes that retain the dead biota. All disturbances remove competitors, but disturbances that remove dead biomass, such as volcanic eruption, fire, or flood, do not produce nutrient-rich conditions.[309] It is mainly the open rocky debris fields at the bases of mountains or cliffs that retain permanently the suitable growing conditions for these plants. These are the "persistent waste places" described by Marks (1983). So even though the fields on prairie terrain are flat, their structure and function is identical to that of an open and heavily disturbed debris field at the base of a cliff or mountain. Instead of rockfall being the agent that continuously supplies the disturbances that reduce competition, we now use tractors, plows and, unfortunately since World War II, herbicides to achieve the same effect.

The ecological literature is filled with references to the highly opportunistic nature of weed species. The ability of weeds to exploit both natural and human disturbances is attributed to a generalist life history strategy that allows them to cope equally well with all environments. The Urban Cliff Hypothesis argues, by contrast, that the organisms that live in urban environments, including humans, are highly specialized to exploit one particular naturally occurring habitat in which opportunism is strongly selected for: cliffs, caves, and other rocky places. To these organisms, human-modified environments are ecologically equivalent to the places where they evolved. In this sense, these species are true endemics. Pigeons on statues think that they are sitting on cliffs near abundant sources of food!

We hope that the negative stereotype of cave man may be replaced with the concept of cliff dweller. This involves more than just the simple substitution of a label. An entire array of pejorative language and erroneous ideas about our ancestors and our ecological history must be replaced with a much more positive image. Instead of viewing our ancestors with a kind of tolerant disgust, we should acknowledge that we still live in environments and show behaviours that would be instantly familiar to them. Our ancestors would certainly be surprised at the *form* of the tools we construct from mineral materials mined from the earth, but they would not be surprised that we make tools from the mineral substrate at all. Equally, they might very well marvel at the extent to which graphic arts, sculpture, and music have changed in form and volume, but they would immediately understand that these forms of expression are central to our humanity. The degree to which we have overcome distance and time as barriers to communication would surprise them, but they were just as interdependent and obligately cooperative with each other as we are today and would instantly recognize that the removal of barriers to communication increases the ability to safely exploit nature. Lastly, if they were invited for dinner, our ancestors would be familiar with most of the plants and animals on the menu. They might be surprised at the social formats while dining – McDonald's-style opportunities for instantaneous foraging probably didn't exist in Paleolithic times – but the food itself has many of the same components, and they might permit themselves a knowing smile at disputes over feeding hierarchies and social customs during family meals.

There is also a darker side to the Urban Cliff Hypothesis. Early hominids regarded their dwellings and the surrounding productive lands as territories to be defended. Competitive interactions would have naturally led to the development of social structures and mechanisms such as theft, assault, rape, and murder that would disadvantage competing groups. Work by Gamble (1999a) suggests that Neanderthals and modern early humans had social networks and population interactions that were organized along the same lines as many, or maybe even most, modern societies. Thus, our inclination to distinguish between 'us' and 'them' is an ancient characteristic that is probably both learned and inherited. As a consequence, our ancestors would recognize not only most of the animals and plants that may be the subjects of disputes, but also other forms of social interaction such as mate competition, dwelling competition, and workplace competition. They would immediately understand the functional significance of most components of our modern urban landscape, and probably would be willing to use force to defend these resources from competitors.

The Urban Cliff Hypothesis suggests that human beings were different from other organisms living in the Pleistocene in one important way. We are not simply a species that has done well at the expense of our environment. What accounts for our enormous success is that we have managed to propagate our entire habitat, including the species of plants and animals that originally shared the habitat with us. Each of these species now has a population size and growth rate that vastly exceeds that of their closely related sibling species living in other habitats. In fact, many of these sibling species are now on the verge of being driven to extinction by the plants and animals that are associated with humans. Quammen (1996) views the success of species such as the rock dove as evidence of a dreary future for our species, mourning the loss of the more diverse, beautiful, and interesting plants and animals that once occupied the land now converted to urban landscape. We argue that we should be amazed at the wondrous adaptability and success of the organisms that were once endemic to rock outcrops and have so consistently associated themselves with us over hundreds of thousands of years. This does not mean that we wish to see a greater conversion of natural landscapes into urban canyons. However, it does mean that we should have a greater respect and affection for these organisms with which we have shared a habitat for the greater part of the Pleistocene epoch.

For those wishing to change the trajectory of the conversion of natural habitats into human-modified or human-built ones all over the world, we believe that the task is daunting. Culture is largely a learned feature of each human population, and abundant evidence shows that legislation and social policy can direct the behaviours of people toward more environmentally sustainable levels. But if our hypothesis is correct, the past million years of biological evolution has included the evolution of our attitudes toward the building of homes or paradises for ourselves and our families. The form of this construction itself has not changed materially in the 10,000 years since the invention of the brick. To achieve meaningful environmental sustainability, somehow this concept of home that has developed over the past million years must be made to accommodate new approaches that are less environmentally costly.

None of this can be done unless people understand where we have come from as a species and where we have been. We have argued that people have forgotten about the paradise that was once available in every dwelling. We hope this book will help generate this understanding. More than that, we hope it will also awaken an appreciation for the myriad of paradises that we still have the power to create on this planet.

---

307 Funk & Wagnall's Standard College Dictionary. 1963. Longman's Canada Limited, Toronto.
308 Berman (1999).
309 However, flood zones do compensate for the losses of organic matter by the deposition of new nutrient-rich sediment from upstream. In fact, some agriculturally important plants such as rice may have evolved from grasses that exploited the seasonal flushes of nutrients into flood zones.

References Cited

Abdul-Rauf, M. 1978. Pilgrimage to Mecca. Nat. Geo. 154, 581–607.

Aigner, J.S. 1978a. Pleistocene faunal and cultural stations in south China. Pages 129–160 in F. Ikawa-Smith (ed.) Early Paleolithic in south and east Asia. The Hague: Mouton Publishers.

Aigner, J.S. 1978b. Important archeological remains from north China. Pages 163–232 in F. Ikawa-Smith (ed.) Early Paleolithic in south and east Asia. The Hague: Mouton Publishers.

Alexander, C. 1977. A pattern language. New York: Oxford University Press.

Anderson, R.C., Fralish, J.S., and Baskin, J.M. 1999. Savannas, barrens and rock outcrop communities of North America. Cambridge: Cambridge University Press.

Ando, T. and Hashimoto, G. 1998. Two new species of *Petunia* (Solanaceae) from southern Rio Grande do Sul, Brazil. Brittonia 50, 483–492.

Andrews, P. 1990. Owls, caves and fossils. Chicago: University of Chicago Press.

Appleton, J. 1975. The experience of landscape. London: Wiley.

Argue, D. 1995. Aboriginal occupation of the southern highlands: was it really seasonal? Australian Archaeology 41, 30–36.

Audubon, J.J. 1989. Audubon's quadrupeds of North America. Wellfleet, NJ: Secacus. Reprint of Audubon, J.J. (1854), The Viviparous quadrupeds of North America, second edition. Philadelphia: Audubon.

Ault, J.P. 1922. Sailing the seven seas in the interest of science. Nat. Geo. 42, 631–690.

Austad, S. 2002. A mouse's tale. Nat. Hist.111, 64–70.

Avery, D.M. 1995. Southern savannas and Pleistocene hominid adaptations: the micromammalian perspective. Pages 459–478 in E.S. Vrba, G.H. Denton, T.C. Partridge, and L.H. Burckle (eds.) Paleoclimate and evolution. New Haven and London: Yale University Press.

Baker, H.G. and Stebbins, G.L. (eds.) 1965. The genetics of colonizing species. New York: Academic Press.

Baker, H.G. 1974. The evolution of weeds. Ann. Rev. Ecol. Syst. 5, 1–24.

Balling, J.D. and Falk, J.H. 1982. Development of visual preferences for natural landscapes. Environment and Behaviour 14, 5–28.

Bar-Yosef, O. 1995. The role of climate in the interpretation of human movements and cultural transformations in western Asia. Pages 507–523 in E.S. Vrba, G.H.

Denton, T.C. Partridge, and L.H. Burckle (eds.) Paleoclimate and evolution. New Haven and London: Yale University Press.

Barigozzi, C. 1986. The origin and domestication of cultivated plants. New York: Elsevier.

Baring-Gould, S. 1911. Cliff castles and cave dwellings of Europe. London: Seeley and Co.

Bassett, M. 1996. Life on the vertical. Nat. Can. Winter 1996, 23–28.

Beadle, G.W. 1977. Origins of *Zea mays*. Pages 615–633 in C.A. Reed (ed.) Origins of Agriculture. The Hague: Mouton Publishers.

Beek, M.A. 1962. Atlas of Mesopotamia. Translated by D.R. Welsh. London: Nelson.

Berman, J.C. 1999. Bad hair days in the Paleolithic: modern (re)constructions of the cave man. Am. Anth. 101, 288–304.

Bernbaum, E. 1997. Sacred mountains of the world. Berkeley: University of California Press.

Binford, L.R. 1984. Faunal remains from the Klasies River Mouth. Orlando: Academic Press.

Binford, L.R. and Ho, C.K. 1985. Taphonomy at a distance: Zhoukoudian, "The cave home of Beijing man"? Curr. Anth. 26, 413–442.

Bingham, P.M. 2000. Human evolution and human history: a complete theory. Evol. Anth. 20, 248–257.

Bird, C.F.M., Frankel, D., and Van Waarden, N. 1988. New radiocarbon determinations from the Grampians-Gariwerd region, western Victoria. Archaeology in Oceania 33, 31–36.

Bogucki, P. 1996. The spread of early farming in Europe. Am. Sci. 84, 242–253.

Bonavia, E. 1890. The cultivated oranges and lemons of India and Ceylon. London: W.H. Allen.

Bordes, F. 1972. A tale of two caves. New York: Harper and Row.

Bourassa, S.C. 1991. The aesthetics of landscape. London: Belhaven Press.

Boyer, P. 2000. Functional origins of religious concepts: ontological and strategic selection in evolved minds. J. Roy. Anth. Inst. 6, 195–214.

Brady, J.E. and Prufer, K.M. 1999. Caves and crystalmancy: evidence for the use of crystals in ancient Maya religion. J. Anth. Res. 55, 129–144.

Brown, C.R. and Brown, M.B. 1995. Cliff swallow. The birds of North America 149, 1–31

Buckler, E.S., Persall, D.M., and Holtsford, T.P. 1998. Climate, plant ecology, and central Mexican archaic subsistence. Curr. Anth. 39, 152–164.

Budiansky, S. 1992. The covenant of the wild. New Haven and London: Yale University Press.

Bunney, S. 1994. Did modern culture begin in prehistoric caves? New Scientist 141, 16.

Butzer, K.W. and Isaac, G. 1975. After the Australopithecines. The Hague, Paris: Mouton Publishers.

Cabe, P.R. 1993. European starling. The birds of North America 48, 1–23.

Caldwell, J.R. 1977. Cultural evolution in the Old World and the New, leading to the beginnings and spread of agriculture. Pages 74–93 in C.A. Reed (ed.) Origins of Agriculture. The Hague: Mouton Publishers.

Cameron, G.G. 1950. Darius carved history on ageless rock. Nat. Geo. 98, 825–844.

Carbonell, E. and Vaquero, M. 1998. Behavioral complexity and biocultural change in Europe around forty thousand years ago. J. Anth. Res. 54, 373–397.

Carney, M. 1995. Peregrines in sight. Seasons 35, 21–25.

Carr, C. 1977. Why didn't the American Indians domesticate sheep? Pages 637–691 in C.A. Reed (ed.) Origins of Agriculture. The Hague: Mouton Publishers.

Carrión, J.S. and Scott, L. 1999. The challenge of pollen analysis in palaeoenvironmental studies of hominid beds: the record from Sterkfontein caves. J. Hum. Evol. 36, 401–408.

Carswell, J., de la Haba, L., Fishbein, S.L., O'Neill, T., and Ramsay, C.R. (eds.) 1981. Splendors of the past. Lost cities of the Ancient World. Washington: National Geographic.

Casey, E.S. 1993. Getting back into place. Bloomington and Indianapolis: Indiana University Press.

Casteret, N. 1948. Lascaux cave, cradle of world art. Nat. Geo. 94, 771–794.

Chang, T. 1976. The origin, evolution, cultivation, disseminaton, and diversification of Asian and African rices. Euphytica 25, 425–441.

Chauvet, J.-M., Deschamps, E.B., and Hillaire, C. 1996. Dawn of art: the Chauvet cave. New York: Abrams.

Clark, D. 1975. A comparison of the late Acheulian industries of Africa and the Middle East. Pages 605–659 in K.W. Butzer and G. Isaac (eds.) After the Australopithecines. The Hague, Paris: Mouton Publishers.

Clark, G. 1977. World prehistory in new perspective. London: Cambridge University Press.

Clottes, J. and Courtin, J. 1994. The cave beneath the sea. New York: Abrams.

Clottes, J. 1995. Les Cavernes de Niaux. Paris: Suil.

Clutton-Brock, J. 1981. Domesticated animals from early times. Cambridge: Cambridge University Press.

Cook, O.F. 1916. Staircase farming of the ancients. Nat. Geo. 29, 474–534.

Cook, J. 1994. Environmentally benign architecture: beyond passive. Pages 125–152 in R. Samuels and D.K. Prasad (eds.) Global warming and the built environment. London: E. & F.N. Spon.

Cooper, G. and Taylor, G. 1996. Paradise transformed. New York: Monacelli Press.

Cooper, G. and Taylor, G. 2000. Gardens for the future. New York: Monacelli Press.

Coppinger, R. and Coppinger, L. 2001. Dogs. New York: Scribner.

Coulson, D. 1999. Ancient art of the Sahara. Nat.Geo. 195, 98–119.

Courbon, P., Chabert, C., Bosted, P., and Lindsley, K. 1989. Atlas of great caves of the world. St. Louis: Cave Books.

Cox, E.H.M. 1945. Plant hunting in China. London: Oldbourne.

Cross, I. 1999. Is music the most important thing we ever did? Music, development and evolution. Pages 10–39 in Suk Won Yi, (ed.) Music, Mind and Science. Seoul: Seoul National University Press.

Cross, I., Zubrow E., and Cowan, F. 2002. Musical behaviours and the archaeological record: a preliminary study. Exp. Archaeol. British Archaeological Reports International Series 1035, 25–34.

Crouch, D.P. 1985. History of architecture. New York: McGraw-Hill.

Dams, L. 1984. Preliminary findings at the "organ" sanctuary in the cave of Nerja, Malaga, Spain. Ox. J. Arch. 3, 1–13.

Dams, L. 1985. Paleolithic lithophones: descriptions and comparisons. Ox. J. Arch. 4, 31–46.

Darlington, A. and Dixon, M. 2000. The biofiltration of indoor air III. Air flux and temperature and removal of VOC's. Proceedings, 2000 USC-TRG conference on biofiltration, USC.

Darlington, A.B., Dat, J.F., and Dixon, M.A. 2001. The biofiltration of indoor air: air flux and temperature influences the removal of toluene, ethylbenzene, and xylene. Env. Sci. Tech. 35, 240–246.

Dart, R.A. 1925. *Australopithecus africanus*: The man-ape of South Africa. Nature 115, 195–199.

Darwin, C. 1859. On the origin of species. London: Murray.

Davis, N. 1981. Human sacrifice. New York: William Morrow & Co.

Davis, P.H. 1951. Cliff vegetation in the eastern Mediterranean. J. Ecol. 39, 63–93.

Deacon, H.J. 1975. Demography, subsistence and culture during the Acheulian in southern Africa. Pages 543–569 in K.W. Butzer and G. Isaac (eds.) After the Australopithecines. The Hague, Paris: Mouton Publishers.

Deacon, J. 1999. South African rock art. Evol. Anth. 8, 48–64.

De Candolle, A. 1964. Origin of cultivated plants. Reprint of second edition 1886. New York and London: Hafner Publishing.

De Lumley, H. 1969. A paleolithic camp at Nice. Sci. Am. 225, 42–59.

De Lumley, H. 1975. Cultural evolution in France in its paleoecological setting during the middle Pleistocene. Pages 745–808 in K.W. Butzer and G. Isaac (eds.) After the Australopithecines. The Hague, Paris: Mouton Publishers.

De Lumley, H. and Darlas, A. 1994. Grotte de Kalamakia (Aréopolis, Péloponnèse). Bulletin de Correspondence Hellenique 118, 535–559.

Delcourt, H. 1987. The impact of pre-historic agriculture and land occupation. Trends in Ecology and Evolution 2, 39–44.

De Turk, P.E. 1968. An introductory history of medieval castle gardens. MLA thesis, Urbana, IL.

Diamond, J. 1992. The third chimpanzee. New York: HarperCollins.

Diamond, J. 1999. Guns, germs and steel. New York: W.W. Norton.

Doebley, J. 1990. Molecular evidence and the evolution of maize. Econ. Bot. 44, 6–27.

Dolling, W.R. 1991. The Hemiptera. New York: Oxford University Press.

Dubos, R. 1965. Man adapting. New Haven: Yale University Press.

Dubos, R. 1968. So human an animal. New York: Scribner.

Easterbee, N., Hepburn, L.V., and Jefferies, D.J. 1991. Survey of the status and distribution of the wildcat in Scotland, 1983–1987. Peterborough: Nature Conservancy Council for Scotland.

Eisenberg, E. 1998. The ecology of Eden. Toronto: Random House.

El Hadidi, M.N., El-Ghani, M.A., Springuel, I. and Hoffman, M.A. 1986. Wild barley *Hordeum spontaneum* L. in Egypt. Biol. Cons. 37, 291–300.

Ellenberg, H. 1988. Vegetation ecology of central Europe. Cambridge: Cambridge University Press.

Fagg, B. 1956. The discovery of multiple rock gongs in Nigeria. Man 56, 17–18.

Fagg, M.C. 1997. Rock Music. Pitt Rivers Museum, University of Oxford, Occasional Papers on Technology No. 14. Oxford: Oxford University Press.

Fairchild, D. 1919. A hunter of plants. Nat. Geo. 36, 57–76.

Farrar, R. 1917. On the eaves of the world. Vols. I and II. London: Edward Arnold.

Feldhammer, G.A., Gates, J.E., and Chapman, J.A. 1984. Rare, threatened, endangered and extirpated mammals from Maryland. Pages 395–438 in A.W. Norden, D.C. Forester, and G.H. Fenwick (eds.) Threatened and endangered plants and animals of Maryland. Maryland Natural Heritage Program Special Publication, 84-I. Towson, MD: Maryland Dept. of Natural Resources.

Fox, R.B. 1978. The Philippine Paleolithic. Pages 59–85 in F. Ikawa-Smith (ed.) Early Paleolithic in south and east Asia. The Hague: Mouton Publishers.

Gamble, C. 1986. The Palaeolithic settlement of Europe. Cambridge: Cambridge University Press.

Gamble, C. 1994. Timewalkers. Cambridge: Harvard University Press.

Gamble, C. 1999a. The Palaeolithic societies of Europe. Cambridge: Cambridge University Press.

Gamble, C. 1999b. Gibraltar and the Neanderthals 1848–1998. J. Hum. Evol. 36, 239–243.

Gargett, R.H. 1999. Middle Palaeolithic burial is not a dead issue: the view from Qafzeh, Saint-Césaire, Kebara, Amud, and Dederiyeh. J. Hum. Evol. 37, 27–90.

Gibson, A.C. and Nobel, P.S. 1986. The cactus primer. Cambridge: Harvard University Press.

Girard, R. 1989. The scapegoat. Baltimore: Johns Hopkins University Press.

Gladkihm, M.I., Kornietz, N.L., and Soffer, O. 1984. Mammoth bone dwellings on the Russian plain. Sci. Am. 251, 164–170.

Golovanova, L.V., Hoeffecker, J.F., Kharitonov, V.M., and Romanova, G.P. 1999. Mezmaiskaya Cave: a Neandertal occupation in the northern Caucasus. Curr. Anth. 40, 77–88.

Goodall, J. 1988. In the shadow of man. Boston: Houghton Mifflin.

Gorman, C. 1977. A priori models and Thai history: a reconsideration of the beginnings of agriculture in southeast Asia. Pages 321–355 in C.A. Reed (ed.) Origins of Agriculture. The Hague: Mouton Publishers.

Govinda, A. 1970. The way of the white cloud. Berkeley: Shambala Press.

Groff, W.G. and Lau, T.C. 1937. Landscaped Kwangsi, China's province of pictorial art. Nat. Geo. 72, 671–710.

Guidoni, E. 1978. Primitive architecture. New York: Abrams.

Han, Y.-C. 1178. Monograph on the oranges of Wen-Chou, Chekiang, translated by M.J.Hagerty, 1923. Leide: E.J. Brill.

Hanelt, P. 1985. Zur Taxonomie, Chorologie und Ökologie der Wildarten von *Allium* L. sect. *cepa* (Mill.) Prokh. Flora 176, 99–116.

Harlan, H. 1925. A caravan journey through Abyssinia. Nat. Geo. 48, 613–663.

Harper, J.L. 1977. The population biology of plants. London: Academic Press.

Harrisson, T. 1978. Present status and problems for Paleolithic studies in Borneo and adjacent islands. Pages 37–57 in F. Ikawa-Smith (ed.) Early Paleolithic in south and east Asia. The Hague: Mouton Publishers.

Harvati, K. and Delson, E. 1999. Conference report: paleoanthropology of the Mani Peninsula (Greece). J. Hum. Evol. 36, 343–348.

Heerwagen, J.H. and Orians, G.H. 1993. Humans, habitats and aesthetics. Pages 138–172 in S.R. Kellert and E.O.Wilson (eds.) The biophilia hypothesis. Washington: Shearwater Books.

Heiser, C.B. 1965. Sunflowers, weeds, and cultivated plants. Pages 391–401 in H.G. Baker and G.L. Stebbins (eds.) The genetics of colonizing species. New York: Academic Press.

Helms, R. 1890. Report of a collecting trip to Mount Kosciusko. Records of the Australian Museum 1, 11–16.

Higham, C.F.W. 1977. Economic change in prehistoric Thailand. Pages 385–412 in C.A. Reed (ed.) Origins of agriculture. The Hague: Mouton Publishers.

Hirsch, E. and O'Hanlon, M. (eds.) 1995. The anthropology of landscape. Oxford: Clarendon Press.

Hoebel, E.A. 1966. Anthropology: the study of man. New York: McGraw-Hill.

Hogg, G. 1958. Cannibalism and human sacrifice. London: Robert Hale.

Hood, M. 1980. The evolution of the Javanese gamelan. Book I. Music of the roaring sea. New York: C.F. Peters.

Howell, F.C. 1965. Early man. New York: Time-Life.

Howell, J.M. 1987. Early farming in northwestern Europe. Sci. Am. 257, 118–126.

Hubbard, A.L., McOrist, S., Jones, T.W., Boid, R., Scott, R., and Easterbee, N. 1992. Is survival of European wildcats *Felis silvestris* in Britain threatened by interbreeding with domestic cats? Biol. Cons. 61, 203–208.

Hull, R.B. and Stewart, W.P. 1992. Validity of photo-based scenic beauty judgments. J. of Env. Psychology 12, 101–114.

Humphrey, C. 1995. Chiefly and shamanist landscapes in Mongolia. Pages 135–162 in E. Hirsch and M.O. Hanlon (eds.) The anthropology of landscape. Oxford: Clarendon Press.

Im, S. 1984. Visual preferences in enclosed urban spaces: an exploration of scientific approaches to environmental design. Environment and Behaviour 16, 235–262.

Isaac, G. 1975. Stratigraphy and cultural patterns in east Africa during the middle ranges of Pleistocene time. Pages 495–542 in K.W. Butzer and G. Isaac (eds.) After the Australopithecines. The Hague, Paris: Mouton Publishers.

Jackson, G. and Sheldon, J. 1949. The vegetation of magnesian limestone cliffs at Markland Grips near Sheffield. J. Ecol. 37, 38–50.

Jackson, G., Gartlan, J.S., and Posnansky, M. 1965. Rock gongs and associated rock paintings on Lolui Island, Lake Victoria, Uganda: a preliminary note. Man 31, 38–40.

Jacobs, J. 1968. The horizon book of great cathedrals. New York: American Heritage Publishing.

Jacobsen, T.W. 1976. 17,000 years of Greek prehistory. Sci. Am. 234, 76–87.

Jaeger, J.-J. 1975. The mammalian faunas and hominid fossils of the middle Pleistocene of the Maghreb. Pages 399–418 in K.W. Butzer and G. Isaac (eds.) After the Australopithecines. The Hague, Paris: Mouton Publishers.

James, E.O. 1965. From cave to cathedral. New York: Praeger.

Jerardino, A. and Swanepoel, N. 1999. Painted slabs from Steenbokfontein Cave: the oldest known parietal art in southern Africa. Curr. Anth. 40, 542–548.

Jessop, L.A. 1921. A pilgrimage to Amernath, Himalayan shrine of the Hindu faith. Nat. Geo. 40, 513–542.

Johanson, D. and Edgar, B. 1996. From Lucy to language. New York: Simon and Schuster.

Johnson, F.E. 1911. The mole-men: an account of the troglodytes of southern Tunisia. Nat. Geo. 22, 787–846.

Johnson, W. 1980. Missouri, the cave state. Jefferson City, MO: Discovery Press.

Johnston, R.F. 1992. Rock dove. The birds of North America 13, 1–14.

Johnston, R.F. and Janiga, M. 1995. Feral pigeons. New York: Oxford University Press.

Jordan, R.P. 1979. Time of testing for an ancient land, Sri Lanka. Nat. Geo. 155, 122–150.

Jordan III, W.R., Gilpin, M.E., and Aber, J.D. 1987. Restoration ecology. Cambridge: Cambridge University Press.

Juniper, B. 2000. Prehistoric pippins. Oxford Today 13, 28–29.

Kahlke, H.D. 1975. The macrofaunas of continental Europe. Pages 309–374 in K.W. Butzer and G. Isaac (eds.) After the Australopithecines. The Hague, Paris: Mouton Publishers.

Kaplan, S. 1992. Environmental preference in a knowledge-seeking, knowledge-using organism. Ch. 16. Pages 581–598 in J.H. Barkow, L. Cosmides and J. Tooby (eds.) The adapted mind. Oxford: Oxford University Press.

Keddy, P.A. 1989. Competition. London: Chapman and Hall.

Kellert, S. 1993. The biological basis for human values of nature. Pages 42–69 in S.R. Kellert and E.O. Wilson (eds.) The biophilia hypothesis. Washington: Island Press, Shearwater Books.

Kellert, S.R. and Wilson, E.O. (eds.) 1993. The biophilia hypothesis. Washington: Island Press, Shearwater Books.

Kelly, P.E., Larson, D.W., and Cook, E.R. 1992. Constrained growth, cambial mortality, and dendrochronology of ancient *Thuja occidentalis* on cliffs of the Niagara Escarpment: an eastern version of Bristlecone Pine? Int. J. Plant Sci. 154, 1117–1127.

Kempe, D. 1988. Living underground. London: Hebert Press.

Kitchener, A.C. and Easterbee, N. 1992. The taxonomic status of black wild felids in Scotland. J. Zoo. Lond. 227, 342–346.

Klein, R.G. 1973. Ice age hunters of the Ukraine. Chicago: University of Chicago Press.

Kostof, S. 1985. A history of architecture. New York: Oxford University Press.

Kraybill, N. 1977. Pre-agricultural tools for the preparation of foods in the Old World. Pages 485-521 in C.A. Reed (ed.) Origins of agriculture. The Hague: Mouton Publishers.

Krings, M., Stone, A., Schnitz, R.W., Krainitski, H., Stoneking, M., and Pääbo, S. 1997. Neandertal DNA sequences and the origin of modern humans. Cell 90, 19–30.

Krings, N., Geisert, H., Schmitz, R.W., Krainitski, H., and Pääbo, S. 1999. DNA sequence of the mitochondrial region II from the Neandertal type specimen. Proc. Nat. Acad. Sci. U.S.A. 96, 5581–5585.

Krings, N., Capelli, C., Tchentscher, F., Geisert, H., Meyer, S.V., Haeseler, A., Grossschmidt, K., Possnert, G., Paunovic, M., and Pääbo, S. 2000. A view of Neandertal genetic diversity. Nat. Genet. 26, 144–146.

Kunzig, R. 1999. Learning to love Neanderthals. Discover 20, 68–75.

Kurtén, B. 1965a. The carnivora of the Palestine caves. Acta Zool. Fennica 107, 1–74.

Kurtén, B. 1965b. On the evolution of the European Wild Cat Felis silvestris Schreiber. Acta Zool. Fennica 111, 1–29.

Kurtén, B. 1976. The cave bear story. New York: Columbia University Press.

Lajoux, J.-D. 1963. The rock paintings of Tassili. London: Thames and Hudson.

Larick, R. and Ciochon, R.L. 1996. The African emergence and early Asian dispersals of the genus Homo. Am. Sci. 86, 538–551.

Larson, D.W., Matthes, U., and Kelly, P.E. 1999. Cliffs as natural refuges. Am. Sci. 87, 410–417.

Larson, D.W., Matthes, U., and Kelly, P.E. 2000a. Cliff ecology. Cambridge: Cambridge University Press.

Larson, D.W., Matthes, U., Gerrath, J.A., Larson, N.W.K., Gerrath, J.M., Nekola, J.C., Walker, G.L., Porembski, S., and Charlton, A. 2000b. Evidence for the widespread occurrence of ancient forests on cliffs. J. Biog. 27, 319-331.

Laurie, M. 1975. An introduction to landscape architecture. New York: American Elsevier.

Lavaud-Girard, F. 1993. Macrofauna from Castelperronian levels at Sainte-Césaire, Charente-Maritime. Pages 73–77 in F. Lévêque, A.M. Backer, and M. Guilbaud (eds.) Context of a late Neandertal. Monographs in World Archeology No. 16. Madison, WI: Prehistory Press.

Leakey, L.S.B. 1960. Adam's ancestors. New York: Harper and Row.

Leakey, M.D. 1983. Tanzania's stone age art. Nat. Geo. 164, 84–99.

Lee, W.T. 1924. A visit to Carlsbad cavern. Nat. Geo. 45, 1–32.

Legge, A.J. 1972. Cave climates. Pages 97–103 in E.S. Higgs (ed.) Papers in economic prehistory. Cambridge: Cambridge University Press.

Lerche, P. 2000. Lost tombs of Peru. Nat. Geo. 198, 64–81.

Lévêque, F. 1993. Introduction to Saint-Césaire. Pages 3–6 in F. Lévêque, A.M. Backer, and M. Guilbaud (eds.) Context of a late Neandertal. Monographs in World Archeology No. 16. Madison, WI: Prehistory Press.

Lewis-Williams, D. 2001. Paintings of the spirit. Nat. Geo. 199, 118–125.

Lewis-Williams, D. 2002. The mind in the cave: consciousness and the origins of art. London: Thames and Hudson.

Liebman, M., Mohler, C.L., and Staver, C.P. 2001. Ecological management of agricultural weeds. Cambridge: Cambridge University Press.

Lieth, H. and Whittaker, R.H. 1975. Primary productivity in the biosphere. Berlin: Springer.

Löve, D. 1992. Translation of Vavilov, N.I (1926). Origin and geography of cultivated plants. Cambridge: Cambridge University Press.

Lowther, P.E. and Cink, C.L. 1992. House sparrow. The birds of North America 12, 1–19.

Lozano, J., Virgós, E., Malo, A.F., Huertas, D.L., and Casanovas, J.G. 2003. Importance of scrub-pastureland mosaics for wild-living cats occurrence in a Mediterranean area: implications for the conservation of the wildcat (*Felis silvestris*). Biodiversity and Conservation 12, 921–935.

Lyell, C. 1873. The antiquity of man. Fourth edition. London: John Murray.

Maass, A., Merici, I., Villafranca, E., Furlani, R., Gaburro, A., Getrevi, A., and Masserini, M. 2000. Intimidating buildings. Can courthouse architecture affect perceived likelihood of conviction? Env. and Behav. 32, 674–683.

Maccoby, H. 1982. The sacred executioner. Bath: Thames and London.

MacNeish, R.S. 1964. The origins of New World civilization. Sci. Am. 211, 29–37.

MacNeish, R.S. 1977. The beginning of agriculture in central Peru. Pages 753–802 in C.A. Reed (ed.) Origins of Agriculture. The Hague: Mouton Publishers.

Marean, C.W. 1998. A critique of the evidence for scavenging by Neandertals and early modern humans: new data from Kobeh Cave (Zagros Mountains, Iran) and Die Kelders Cave 1, Layer 10 (South Africa). J. Human Evol. 35, 111–136.

Marks, P. 1983. On the origin of the field plants of the northeastern United States. Am. Nat. 122, 210–227.

Marshack, A. 1975. Exploring the mind of ice age man. Nat. Geo. 147, 64–89.

Matthes-Sears, U., Gerrath, J., and Larson, D.W. 1997. Abundance, biomass, and productivity of endolithic and epilithic lower plants on the temperate-zone cliffs of the Niagara Escarpment, Canada. Int. J. Plant Sci. 158, 541–560.

McGrew, W.C., McKee, J.K., and Tutin, C.E.G. 2003. Primates in caves: two new reports of *Papio* spp. J. Human Evol. 44, 521–526.

McNatt, L. 1996. Cave archaeology of Belize. Journal of Cave and Karst Studies 58, 81–99.

McNeill, W.H. 1976. Plagues and peoples. Garden City: Anchor Press.

Mellart, J. 1964. A Neolithic city in Turkey. Sci. Am. 210, 94–104.

Mercader, J., Garralda, M.D., Pearson, O.M., and Bailey, R.C. 2001. Eight-hundred-year-old human remains from the Ituri tropical forest, Democratic Republic of the Congo: the rock shelter site of Matangai Turu Northwest. Am. J. Phys. Anth. 115, 24–37.

Meyer, P.J. 1969. Foxes foretell the future in Mali's Dogon country. Nat. Geo. 135, 431–448.

Milton, K. 1999. A hypothesis to explain the role of meat-eating in human evolution. Evol. Anth. 8, 11–21.

Mitchell, P.J., Yates, R., and Parkington, J.E. 1996. At the transition in southern Africa. Pages 15–36 in L.G. Straus, B.V. Eriksen, J.M. Erlandson, and D.R.Yesner (eds.) Humans at the end of the Ice Age. New York: Plenum Press.

Mithen, S.J. 1990. Thoughtful foragers. Cambridge: Cambridge University Press.

Monteith, J.L. and Unsworth, M.K. 1990. Principles of environmental physics. Second edition. London: Edward Arnold.

Moore, R.W. 1960. Angkor, jewel of the jungle. Nat. Geo. 117, 517–569.

Morey, D. 1994. The early evolution of the domestic dog. Am. Sci. 82, 336–347.

Morrish, W.R. 1996. Civilizing terrains, mountains, mounds, and mesas. San Francisco: William Stout.

Mulliken, M.A. 1938. China's great wall of sculpture. Nat. Geo. 73, 313–348.

Naveh, Z. and Vernet, J.-L. 1991. The palaeohistory of the Mediterranean biota. Pages 19–32 in R.H. Groves and F.D. Castri (eds.) Biogeography of Mediterranean invasions. Cambridge: Cambridge University Press.

Nesse, R.M. and Williams, G.C. 1996. Why we get sick. New York: Vintage Books.

Nowak, R.M. and Paradiso, J.L. 1983. Walker's mammals of the world. Baltimore: Johns Hopkins University Press.

Nuttgens, P. 1997. The story of architecture. Second edition. London: Phaidon Press.

O'Connell, J.W. and Korff, A. 1991. The book of the burren. Galway, Ireland: Eolas.

Orians, G. 1986. An ecological and evolutionary approach to landscape aesthetics. Pages 3–25 in E.C. Penning-Rowsell and D. Lowenthal (eds.) Landscape meanings and values. London: Allen and Unwin.

Orians, G. and Heerwagen, J.H. 1992. Evolved responses to landscape. Pages 555–579 in J.H. Barkow, L. Cosmides, and J. Tooby (eds.) The adapted mind. Oxford: Oxford University Press.

O'Sullivan, P. 1994. Energy and architectural form. Pages 113–124 in R. Samuels and D.K. Prasad (eds.) Global warming and the built environment. London: E. and F.N. Spon.

Otte, M., Yalçinkaya, A., Kozlowski, J., Bar-Yosef, O., López Bayón, I., and Taskiran, H. 1998. Long-term technical evolution and human remains in the Anatolian Palaeolithic. J. Hum. Evol. 34, 413–431.

Ovchinnikov, T.V., Götherström, A., Romanova, G.P., Kharitonov, V.M., Liden, K., and Goodwin, W. 2000. Molecular analysis of Neanderthal DNA from the northern Caucasus. Nature 404, 490–493.

Patou-Mathis, M. 1993. Taphonomic and paleoethnographic study of the fauna associated with the Neandertal of Saint-Césaire. Pages 81–102 in F. Lévêque, A.M. Backer, and M. Guilbaud (eds.) Context of a late Neandertal. Monographs in World Archeology No. 16. Madison, WI: Prehistory Press.

Pavord, A. 1999. The Tulip. New York: Bloomsbury.

Perkins, E. 1909. With the monks at Meteora: the monasteries of Thessaly. Nat. Geo. 20, 799–807.

Peterken, G.F. 1996. Natural woodland. Cambridge: Cambridge University Press.

Pfeiffer, J. 1972. The emergence of man. New York: Harper and Row.

Pigott, S. 1965. Ancient Europe. London: Aldine Press.

Pimental, D., Lach, L., Zuniga, R., and Morrison, D. 2000. Environmental and economic costs of nonindigenous species in the United States. Bioscience 50, 53–63.

Ping-Ti, H. 1977. The indigenous origins of Chinese agriculture. Pages 413–484 in C.A. Reed (ed.) Origins of Agriculture. The Hague: Mouton Publishers.

Pokines, J.T. 2000. Microfaunal research design in the Cantabrian Spanish Paleolithic. J. Anth. Res. 56, 95–112.

Posnansky, M. 1980. An archeological reconnaissance of Togo. Archeology at UCLA 2, 1–4.

Potts, R. 1984. Home bases and early hominids. Am. Sci. 72, 338–347.

Prasad, K.N. 1996. Pleistocene cave fauna from peninsular India. J. Caves and Karst Studies 58, 30–34.

Pringle, H. 1988. Boneyard enigma. Equinox 38, 87–103.

Pysek, P., Prach, K, and Wade, M. (eds.) 1995. Plant invasions. Amsterdam: Academic Publishing.

Quammen, D. 1996. Superdove on 46th Street. Outside 21, 33–38.

Quammen, D. 1998. Planet of weeds. Harper's Magazine, October 1998, 57–69.

Quennell, P. 1971. The Colosseum. New York: Newsweek Book Division.

Ragir, S., Rosenberg, M., and Tierno, P. 2000. Gut morphology and the avoidance of carrion among chimpanzees, baboons, and early hominids. J. Anth. Res. 56, 477–512.

Ranck, G.L. 1968. The rodents of Libya. Smithsonian Inst. Publ. #275. Washington, D.C.: Smithsonian Inst.

Randi, C. and Ragni, B. 1991. Genetic variability and biochemical systematics of domestic and wild cat populations (*Felis silvestris*: Felidae). J. Mamm. 72, 79–88.

Reed, C.A. 1977a. A model for the origin of agriculture in the Near East. Pages 543–567 in C.A. Reed (ed.) Origins of Agriculture. The Hague: Mouton Publishers.

Reed, C.A. (ed.) 1977b. Origins of Agriculture. The Hague: Mouton Publishers.

Rhoades, R.E. 1993. The golden grain: corn. Nat. Geo. 183, 92–117.

Ribe, R.G. 1989. The aesthetics of forestry: what has empirical preference research taught us? Env. Manage. 13, 55–74.

Rock, J.F. 1930. The glories of the Minya Konka. Nat. Geo. 58, 385–437.

Rock, J.F. 1931. Konsa Risumgongba, holy mountain of the outlaws. Nat. Geo. 60, 1–65.

Routledge, S. 1921. The mystery of Easter Island. Nat. Geo. 40, 629–646.

Rudofsky, B. 1977. The prodigious builders. New York: Harcourt Brace Jovanovich.

Rust, M.K., Owens, J.M. and Reierson, D.A. 1995. Understanding and controlling the German cockroach. Oxford: Oxford University Press.

Ryan, W. and Pitman, W. 1998. Noah's flood. New York: Simon and Schuster.

Salvador, R.J. 1997. Maize. The maize page, Encyclopedia of Culture and Society of Mexico. http://maize.agron.iastate.edu/maizarticle.html.

Samuels, R. and Prasad, D.K. 1994. Global warming and the built environment. London: E. and F.N. Spon.

Sandars, N.K. 1972. The epic of Gilgamesh. London: Penguin Books.

Sanderson, E.W., Jaiteh, M., Levy, M.A., Redford, K.H., Wannebo, A.V., and Woolmer, G. 2002. The human footprint and the last of the wild. Bioscience 52, 891–904.

Sanjur, O.I., Piperno, D.R., Andres, T.C., and Wessel-Beaver, L. 2002. Phylogenetic relationships among domesticated and wild species of *Cucurbita* (Cucurbitaceae) inferred from a mitchondrial gene: implications for crop plant evolution and areas of origin. P.N.A.S. 99, 535–540.

Sankalia, H.D. 1978. The early Paleolithic in India and Pakistan. Pages 97–127 in F. Ikawa-Smith (ed.) Early Paleolithic in south and east Asia. The Hague: Mouton Publishers.

Sauer, J.D. 1993. Historical geography of crop plants: a selected roster. Ann Arbor: CRC Press.

Schaeffer, F.A. 1930. A new alphabet of the ancients is unearthed. Nat. Geo. 58, 477–516.

Schepartz, L.A., Miller-Antonio, S., and Bakken, D.A. 2002. Upland resources and the early Palaeolithic occupation of southern China, Vietnam, Laos, Thailand and Burma. World Archaeology 32, 1–13.

Schmidt-Nielsen, K. 1975. Animal Physiology. London: Cambridge University Press.

Schneider, D. 1990. Starling Wars. Nature Canada 19, 33–39.

Semilo, O., Passarino, G., Oefner, P.J., Lin, A.A., Arbuzova, S., Beckman, L.E., De Benedictis, G., Francalacci, P., Kouvatsi, A., Limborska, S., Marcikiae, M., Mika, A., Mika, B., Primorac, D., Silvana Santachiara-Benerecetti, A., Luca Cavalli-Sforza, L., and Underhill, P.A. 2000. The genetic legacy of paleolithic *Homo sapiens sapiens* in extant Europeans: a Y chromosome perspective. Science 290, 1155–1159.

Sherratt, A.(ed.).1980. The Cambridge encyclopedia of archaeology. Cambridge: Cambridge University Press.

Shippee, R. 1934. A forgotten valley in Peru. Nat. Geo. 62, 110–132.

Shor, F. and Shor, J. 1951. The caves of the thousand Buddhas. Nat. Geo. 99, 383–415.

Shreeve, J. 1995. The Neandertal enigma. New York: William Morrow.

Sikkink, L. and Choque, B. 2000. Landscape, gender and community: Andean mountain stories. Anth. Quart. 72, 167–172.

Simmons, N.W. 1976. Evolution of crop plants. New York: Longman.

Simms, E. 1979. The public life of the street pigeon. London: Hutchinson of London.

Simonds, J.O. 1961. Landscape architecture. New York: McGraw-Hill.

Simpich, F. 1930. The giant that is New York. Nat. Geo. 58, 517–583.

Sink, K.C. (ed.). 1984. Petunia. Berlin: Springer.

Smith, B. 1997. The initial domestication of *Cucurbita pepo* in the Americas 10,000 years ago. Science 276, 932–934.

Smith, C. 1992. Late stone age hunters of the British Isles. London and New York: Routledge.

Smith, F.H., Gaines, J.B., and Krusko, N.A. 1999. A juvenile human frontal bone from the French Upper Palaeolithic site of Lacave: significance and problems of interpretation. Int. J. Osteoarch. 9, 237–243.

Solecki, R.S. 1963. Prehistory in Shanidar Valley, Northern Iraq. Science 139, 179–193.

Solomon, A. 1996. Rock art in southern Africa. Sci. Am. 275, 106–113.

Spencer, J.W. and Kirby, K.J. 1992. An inventory of ancient woodland for England and Wales. Biol. Cons. 62, 77–93.

Spirn, S.A. 1984. The granite garden: urban nature and human design. New York: Basic Books.

Spirn, S.A. 1998. The language of landscape. New Haven: Yale University Press.

Stace, C. 1999. Field flora of the British Isles. Cambridge: Cambridge University Press.

Stahle, D.W. and Chaney, P.L. 1994. A predictive model for the location of ancient forests. Nat. Areas. J. 14, 151–158.

Stevens, L. 1991. Genetics and evolution of the domestic fowl. New York: Cambridge University Press.

Stoffle, R.W., Loendorf, L., Austin, D.E., Halmo, D.B., and Bulletts, A. 2000. Ghost dancing in the Grand Canyon. Curr. Anth. 41, 11–32.

Stringer, C. and McKie, R. 1996. African exodus. New York: Henry Holt.

Stuart, G.E. 1981. Maya art treasures. Nat. Geo. 160, 220–235.

Swisher III, C.C., Curtis, G.H., and Lewin, R. 2000. Java Man. New York: Scribner.

Tattersall, I. 1995. The last Neandertal. New York: Macmillan.

Tattersall, I. 1998. Becoming human. San Diego: Harcourt Brace.

Thorpe, I.J. 1996. The origins of agriculture in Europe. London: Routledge.

Tilley, C. 1994. A phenomenology of landscape. Oxford: Berg.

Tilley, C. and Bennett, W. 2001. An archaeology of supernatural places: the case of West Penwith. J. Roy. Anth. Inst. 7, 335–362.

Tilley, C., Hamilton, S., and Bender, B. 2000. Art and the re-presentation of the past. J. Roy. Anth. Inst. 6, 35–62.

Toy, S. 1985. Castles, their construction and history. New York: Dover Press.

Trachenberg, M. and Hyman, I. 1986. Architecture. Englewood Cliffs, N.J. and New York: Abrams.

Trut, L. 1999. Early canid domestication: the farm-fox experiment. Am. Sci. 87, 160–169.

Tuan, Y.-F. 1974. Topophilia. Englewood Cliffs, N.J.: Prentice-Hall.

Turner, A.R. 1966. The vision of landscape in Renaissance Italy. Princeton: Princeton University Press.

Ulrich, R.S. 1993. Biophilia, biophobia, and natural landscapes. Pages 74–143 in S.R. Kellert and E.O.Wilson (eds.) The biophilia hypothesis. Washington: Island Press.

Ursic, K.A., Kenkel, N.C., and Larson, D.W. 1997. Revegetation dynamics of cliff faces in abandoned limestone quarries. J. Appl. Ecol. 34, 289–303.

Usher, M.B. 1979. Natural communities of plants and animals in disused quarries. J. Env. Manage. 8, 223–236.

Van Schaik, C.P., Deaner, R.O., and Merrill, M.Y. 1999. The conditions for tool use in primates: implications for the evolution of material culture. J. Hum. Evol. 36, 719–741.

Vavilov, N.I. 1926. See Löve (1992).

Vilà, C. and Wayne, R.K. 1999. Hybridization between wolves and dogs. Cons. Biol. 13, 195–198.

Villa, P. and Soressi, M. 2000. Stone tools in carnivore sites: the case of Bois Roche. J. Anth. Res. 56, 187–215.

Von Puttkamer, W.J. 1979. Stone age present meets stone age past. Nat. Geo. 155, 60–83.

Vrba, E.S., Denton, G.H., Partridge, T.C., and Burckle, L.H. (eds.) 1995. Paleoclimate and evolution. New Haven and London: Yale University Press.

Waddell, J., O'Connell, J.W., and Korff, A. 1994. The book of Aran. Galway, Ireland: Eolas.

Ward, P. 2002. Future evolution. San Francisco: Freeman.

Wenke, R.J. 1990. Patterns in prehistory. Third edition. Oxford: Oxford University Press.

Wentzel, V. 1953. India's sculpted temple caves. Nat. Geo. 103, 665–678.

White, E. and Brown, D. 1973. The first men. New York: Time-Life.

Wiersema, J.H. and Leon, B. 1999. World economic plants: a standard reference. New York: CRC Press.

Williams, M.O. 1930. New Greece, the centenarian, forges ahead. Nat. Geo. 58, 683–721.

Wilson, E.O. 1984. Biophilia. Cambridge: Harvard University Press.

Wilson, H.V.P. 1946. Geraniums and Pelargonium. New York: Barrows.

Wing, E.S. 1977. Animal domestication in the Andes. Pages 837–859 in C.A. Reed (ed.) Origins of agriculture. The Hague: Mouton Publishers.

Wright, Jr., H.E. 1977. Environmental change and the origin of agriculture in the Old and New Worlds. Pages 281–318 in C.A. Reed (ed.) Origins of agriculture. The Hague: Mouton Publishers.

Wu, Z. and Raven, P. 1996. Flora of China, Volume 15. Beijing: Science Press.

Zeuner, F.E. 1963. A history of domesticated animals. New York: Harper and Row.

Zohary, D. and Hopf, M. 1993. Domestication of plants in the Old World. Oxford: Clarendon Press.

Zube, E.H, Brush, R.O., and Fabos, J.G. 1975. Landscape assessment: values perceptions and resources. Stroudsberg, PA: Hutchinson and Ross.

Index

The Nature Center at W.W. Knight Preserve
29530 White Road
Perrysburg, OH 43551